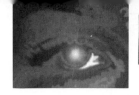
530.76

Synoptic skills in

ADVANCED

PHYSICS

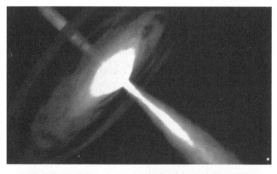

DAVID HOMER

Hodder & Stoughto

A MEMBER OF THE HODDER HEADLINE GR

Acknowledgements

I would like to thank AQA, WJEC and OCR for permission to reproduce adapted versions of their questions. I also want to thank Duncan Dewar, Marc Hull, Michael Speake, Steven Waters and Roland Wilkinson for their help.

Dedication
To Brenda

Orders: please contact Bookpoint Ltd, 130 Milton Park, Abingdon, Oxon OX14 4SB. Telephone: (44) 01235 827720. Fax: (44) 01235 400454. Lines are open from 9.00–6.00, Monday to Saturday, with a 24 hour message answering service. Email address: orders@bookpoint.co.uk

British Library Cataloguing in Publication Data
A catalogue record for this title is available from the British Library

ISBN 0 340 84755 7

First Published 2002
Impression number   10 9 8 7 6 5 4 3 2 1
Year   2007 2006 2005 2004 2003 2002

Typeset by J&L Composition Ltd, Filey, North Yorkshire
Printed in Great Britain for Hodder & Stoughton Educational, a division of Hodder Headline Plc, 338 Euston Road, London NW1 3BH by Martins The Printers, Berwick-upon-Tweed.

CONTENTS

You can find specific advice quickly on the following pages:

Use the rest of this page to note page numbers you use frequently.

CHAPTER ONE

INTRODUCTION

The book is about making connections in physics. Examinations at Advanced level test your synoptic skills – the knowledge and skills that you learn and practise throughout an AS and A2 course. You need to connect these skills and ideas, to master the synoptic skills, if you are to score good marks in your exam.

This book helps you develop your synoptic skills – your ability to connect – in A-level Physics. It offers advice on good study and revision techniques and is designed to help you both during the course and when it is time to revise. There are explanations of the types of synoptic tests you can expect, together with discussions of the various skills you need to develop during your course. Finally, there is a selection of synoptic test questions for you to practise your skills.

GCSE to Advanced level

If you are some way into your A-level courses, you already know that there are many differences between how you study now and what you did up to GCSE. The number of subjects you study has changed, as does the amount of time you devote to each subject over one week. The amount of material covered in each subject is much greater than that at GCSE – and it is taught to you over a smaller time span, perhaps fifty weeks or so in total. You need to be well organised to get the most out of these important courses!

Your learning styles may need to change. A-level students are expected to be far more independent in their learning than people at GCSE – even though there are only a few months between taking GCSEs and starting A-level. This book will help you with this transition to management of your own learning.

AS and A2

Modern A-level students study an AS specification in the first year of the course ('specification' is the name given to the list of topics you have to study and learn). The second-year specification is called A2. Both the AS and the A2 halves of the course make up a full A-level qualification.

The whole set of examinations can be taken in a series of ways. Some candidates will take the six Units for the full A-level one or two at a time throughout the two years. Others will take the full examination in two sections, AS at the end of the first year, A2 at the end of the second year. Some students take all the Units in one go, AS and A2 together at the end of the complete course. This is usually a decision made by your teachers.

AS exams

AS examinations are designed to have a standard somewhere between GCSE and A2 level, a sort of 'half-way house'. The AS questions also have a different emphasis from the later A2 tests (Table 1.1).

Assessment objectives at A-level	
AO1	Knowledge with understanding
AO2	Application of knowledge and understanding, synthesis and evaluation
AO3	Experiment and investigation
AO4	Synthesis of knowledge, understanding and skills **Only tested in the A2 examination**

Table 1.1

The AS examinations emphasise how well you know and understand the material (AO1). This accounts for about 45% of all the marks at AS (awarding bodies are allowed some flexibility in these percentages so your exam may be slightly different). You are also tested on your ability to apply your knowledge (AO2), and these AO2 marks are worth roughly 35%. You may not find the *style* of the question types very different from GCSE. In terms of *content*, however, the questions will change. They may be based on different and more advanced topics or they may contain similar topics to GCSE set at a somewhat higher level of understanding.

The remaining marks (20%) go towards testing your experimental and investigative skills (AO3). This may be tested in a variety of ways, including coursework (assessed by your teachers), practical examinations set by the awarding body (a more recent name for 'examination board') or some other data-analysis tasks.

A2 examinations

An important difference between AS and A2 papers is the inclusion of questions that test *synoptic skills*. You may not have met this type of question at GCSE. The skills themselves will be discussed in detail later and you will be introduced to the types of physics question that can test the skills. A major purpose of this book is to support your physics teacher in teaching these synoptic skills to you.

At A2 the whole of your marks (AS + A2) go to make up the final total. The marks derived from synoptic-skill questions make up 20% of the *whole* mark. This means that, because there is no synoptic testing at AS, about 40% of the A2 marks come from synoptic questions. It is obviously important to score well in this area of the examination and that is where this book is designed to help.

There are other important differences too: synoptic papers need to be taken towards the end of the course. This makes sense because the end of the course is the first time that you will have a complete overview of all the physics you have learnt. (Incidentally, this does not mean that you have to take synoptic papers again if you want to re-sit earlier modules, you are allowed to keep your marks from the final synoptic papers if you want to.)

Summary of the differences between AS and A2

● **The standard of the AS Units is about halfway between GCSE and full A-level.**

● **Most AS marks are for your knowledge with understanding, fewer marks are allocated to your ability to apply your knowledge on the topics in the Unit.**

● **AS contains no synoptic testing. You are only examined on the physics contained in the particular Unit that the examination paper tests.**

● **40% of A2 marks are for synopsis.**

● **You can expect questions covering all parts of the specification in any A2 unit that includes synopsis.**

Using this book for learning

This book gives you advice on how to help your brain learn most effectively. It also suggests how to alter your study methods to suit the new demands of your A-level courses. If possible, read it early in your course and put those ideas that work for you into practice – every day.

Perhaps the most crucial piece of advice for someone who is starting or who is some way into any A-level course is: *think about your study skills*. Sound, robust learning is the real key to both *understanding* physics and being able to *use* your physics under test conditions. Try out new methods of learning, remembering that different methods work for different people. Experiment with different forms of note-taking and other aspects of your study techniques.

Don't just think about study skills at the beginning of your course. Review your learning and your learning styles often, not just the nuts and bolts of the physics, but the *way* you learn. Are there some learning methods that work really well for you? Could they be taken further? Are there things that do not work so well? Have bad learning habits been creeping in?

So, from time to time read and re-read the sections of this book that deal with learning. Try to put into practice the learning advice and the synoptic skills as your own understanding develops.

As you progress further into your courses, look at the specification of the examination that you are taking. This is published by the awarding body. You can obtain a copy from your teacher or you can find it freely available on the Web (the web addresses are at the end of this chapter). Make sure that you understand what each of your examination papers will test – both physics content and style of questions. Be clear (for A2) where and how synoptic skills will be tested (there is help with this in Chapter 2).

Using this book for revision

Start your revision early. Begin by reading the sections in this book that deal with revision. Take the advice given in the book. You should give as much thought to your revision strategies as to your day-to-day learning skills. In particular, if it helps you, draw up an early and realistic revision timetable.

This book contains synoptic questions that test both written and practical skills. Answer these questions and do not be afraid to re-work questions that you answered in the past. Complete questions against the clock only later in your revision schedule. You may take longer to get synoptic answers to the same standard as those set on specific, non-synoptic parts of the specification. It takes time to work your skills up to an examination standard.

Key points for good learning

● Think about and develop good study skills, but do not expect them to develop overnight.

● Know which one of the physics A-level examinations you will be entered for.

● Make sure you know which of your A2 papers test synoptic skills.

● Study your A-level specification as well as your physics.

● Work at examination questions early in your revision schedule, but take it slowly at first.

● Analyse past and specimen papers so that you know how synoptic skills are tested by your awarding body.

Examination boards

AQA
www.aqa.org.uk

EDEXCEL
www.ecexcel.org.uk

OCR
www.ocr.org.uk

WJEC
www.wjec.org.uk

CHAPTER TWO

SYNOPTIC SKILLS IN YOUR EXAMINATION

What is synopsis?

By now you should be asking the question: exactly what do the terms *synopsis* and *synoptic skills* mean?

The awarding bodies make this clear in a series of statements that describe synopsis:

Students should be able to bring together principles and concepts from different areas of physics and apply them in a particular context.

Students should be able to relate their understanding of ideas and skills of physics to empirical data and information in contexts that may be new to them.

Students should be able to use the skills of physics in contexts that bring together different areas of the subject.

The examination papers you sit towards the end of your course will contain some elements of synopsis and will test the physics that you have been learning throughout the course. The exact nature of these tests depends on the particular examination that your centre chooses to use. But whichever set of papers you take, at the end of the course you will need to have an understanding of, and an ability to use, the physics you were taught at the start of the AS year.

For some awarding bodies, the synoptic papers include an element of practical testing, whether through practical examinations or through longer term investigations that you cover at school or college.

So, why is there are need for your synoptic skills to be tested in these ways? An explanation can be found if you think about the differences between vectors and scalars. This is something that you may have learnt early in your career in the context of mechanics. Vectors have a size and a direction; scalars have size only. Velocity – a common example of a vector – always needs a statement of the *direction* in which an object is moving together with a *size* telling you how fast it is going. Speed – just a number, *distance divided by time* – is, on the other hand, a scalar quantity.

The vector–scalar difference goes way beyond mechanics, however. There are vector quantities in all areas of the subject: electricity, magnetism, waves and so on. If you make use of the vector idea in more than one area of the subject whilst you are learning, it will actually enhance your appreciation of the vector–scalar distinction. You have used the idea in a number of contexts.

Without synoptic assessment, students might be tempted to learn the vector idea once, apply it only to mechanics, and then either forget about the idea all together or simply use it in the mechanics context. Synopsis will help learning and it makes students into more complete physicists.

Another example is that of a *field*. Most physicists meet simple ideas of fields in physics early on in GCSE when they learn for the first time about the properties of magnetic fields. They learn the idea of action at a distance, how two magnetic poles can repel or attract each other, and how physicists describe the magnetic force near a magnet in terms of magnetic field lines. Later, beyond GCSE, the field idea is developed to include electric fields and gravitational fields. Finally, there is the merging of electric and magnetic field ideas in the study of the electromagnetic field. The repeated exposure to the field idea constantly develops and expands your understanding of the concept of *field* and makes it easier for you to apply your ideas to other types of field later on.

In synoptic tests you can expect to take ideas and information you already understand from several branches of physics and apply them in contexts (examples) that are new to you.

Some of your physics skills are synoptic too. Skills that you develop in practical work – drawing graphs, treating errors, establishing relationships between variables and so on – spill over into your work in the theory papers. These skills apply to the whole of physics; they are synoptic.

Watch a good driver driving a car through a town. The progress is smooth, safe and unruffled, a remarkable feat given that our brains did not evolve to drive motor cars. Humans were not designed to control a ton of metal moving at speeds of tens of metres every second. But, even more remarkable, is that the apparently effortless interaction of gears, steering and brakes was not, once upon a time, second nature to the driver. The driver had to learn to steer, to use the gear lever in conjunction with the clutch and to release the hand brake at the right moment. And it was probably not always quite so smooth in the early days!

Physics is rather like driving the car. Professional physicists draw together many strands, skills and ideas in the course of their work; these may well come from many areas of the subject. They are synoptic skills. These are what is examined at the end of an A-level course.

So, how do you improve and refine your performance in these synoptic examinations? Like the skills needed to drive the car well, you need to learn your synoptic skills gradually, beginning with simple cases and questions. You should also refine your learning strategies to develop skills that will improve your handling of synoptic questions. These important skills will enhance not just your performance on synoptic papers but also your performance on other, non-synoptic material and your general qualities as a physicist.

Begin by identifying the individual skills of synopsis and using these in your everyday study tasks. At the same time you will be learning new physics ideas. As your understanding of the new concepts builds up, you should apply a synoptic approach in order to link the new ideas together with those that you learnt earlier in your school career. In this way, both new and old ideas are strengthened and refined.

A good A-level physicist – which is what you eventually want to become – will be able to bring the skills developed during the course to bear in any area of the subject, whether studied before or not.

Skills needed by Advanced-level physicists

Theoretical skills

A-level physicists should be able to:

★ recognise, remember and understand:
  – facts (e.g. that velocity is a vector)
  – terms (e.g. what is meant by momentum)
  – principles (e.g. the conservation of energy)
  – relationships (there are some relationships to learn, such as *potential difference = current × resistance*)
  – concepts (e.g. what is meant by a field)
  – practical techniques (how to read a micrometer screw gauge correctly)

★ select, organise and present information logically and clearly

★ interpret data and translate data from one form to another (the data could be a written passage, a table, a diagram or a graph)

★ describe, interpret and explain physical effects and phenomena

★ carry out calculations

★ apply physical principles to unfamiliar situations

★ assess the validity of physical information, experiments, inferences and statements

★ understand the ethical, social, economic, environmental and technological implications and applications of physics.

Practical skills

A-level physicists should be able to:

★ devise and plan experiments that are safe and demonstrate skilful practical techniques

★ take observations and measurements with appropriate precision and record these methodically

★ interpret, explain and evaluate the results of experimental activities.

Mathematical skills

There are some mathematical skills that you need in order to carry out calculations and to understand some parts of the specifications set by the awarding bodies. You can find details of these later in the book.

Communication skills

You are expected to be able to express yourself clearly, in good English. There are usually some marks in written papers assigned to this aspect of your work in the examination. There are some comments about this later in the book.

How your awarding body tests synoptic skills

All the awarding bodies are obliged to use a common framework for the award of marks and for the general standards of A-level examinations, both AS and A2. The bodies do, however, have considerable latitude in the way that they can put together the various parts of the physics examination.

In this section the seven separate examinations for A2 physics provided by the four awarding bodies are considered separately. You need to know for which examination you will be entered so that you can tailor the remainder of the book to your own needs.

Remember that, whichever examination you are entered for, synoptic skills will account for 40% of your A2 marks (20% of the marks for the full A-level qualification because the AS marks count for half of the total).

Types of question

The following section contains specific details of the types of synoptic question set in the seven examinations, but before reading it you need to understand the terms that examiners use to describe the different styles of examination question.

Comprehension – called 'passage analysis' in some specifications

The question contains a short passage that you study for a few minutes. Questions are set on the physics within the passage. Sometimes this is combined with a data-analysis test.

The questions might ask you to explain the meaning of terms used in the passage, to perform calculations and deductions, and may include additional material which has to be related to the content of the passage itself.

Data analysis

The question contains a collection of data in the form of tables, graphs or a written passage. You will be asked to manipulate the data in various ways and to explore the algebraic relationships between variables.

Essay question or **Extended writing**

A question that asks you to explain or describe either a physical effect or an experiment in a piece of continuous writing. There may be marks available for the quality of your written communication as well as the accuracy and completeness of your physics.

Structured question – called *long structured* questions in some specifications

A question that may have several parts all focused around one area of the specification. This type of question has many forms including:

★ *long calculations* split up into parts to lead you towards the answer

★ *short written answers* that separately explore different aspects of a single complex idea

★ *mixtures* of the two types above.

Pre-release comprehension

This type of question is only used in OCR B (Advancing Physics). You will receive a passage about six weeks before the examination date and you will need to study the passage and the physics contained in it. Questions will be set in the context of the passage. The questions in the pre-release comprehension may be in any of the forms above.

Each of the awarding bodies in England has two physics specifications that reflect different styles of teaching (for example, the project-based teaching programmes: *Advancing Physics* and the *Salters Horners Physics Course* are examined by OCR specification B and Edexcel specification B, respectively).

Below are listed the arrangements that each body makes for the synoptic parts of its physics examinations.

AQA (A)

Units with synoptic assessment	Title of unit	Percentage of *whole* A-level qualification
Unit 10 Paper time: 120 minutes	Synoptic	20% (80 marks available)

Table 2.1

This paper consists of structured questions and examines Modules 1–5. It constitutes the whole synoptic assessment for the specification. All questions are compulsory.

AQA (B)

Units with synoptic assessment	Title of unit	Percentage of *whole* A-level qualification	Percentage of this Unit test that is synoptic
Unit 5 Paper time: 120 minutes	Fields and their applications	20%	75%
Unit 6 Paper time: 180 minutes in two separate 90-minute papers	Experimental work	15%	About 33%

Table 2.2

Unit 5

Unit 5 contains both non-synoptic and synoptic questions.

Some of the synoptic questions are based on a comprehension question.

The first questions on the paper will assess your understanding of Module 5 only. Later questions will test all the AS and A2 theory modules (that is, Modules 1, 2, 4 and 5)

Unit 6

Experimental work

There are two parts to this paper.

The first part is taken at a time chosen by your examination centre (usually your own school or college) some time before the main A2 examination session. This paper consists of one or two questions. You will receive a briefing sheet on this part of the Unit before the examination date.

For each question you will be given apparatus and instructions for the experiment you are to perform. For example, you might be told which independent variable you are to investigate, or you might be invited to test a specified hypothesis (theory). You will then develop the method and collect data. You will analyse these data – possibly graphically – and then decide whether the hypothesis is supported or not. In this part of the examination the emphasis is on your experimental skills of:

★ developing and implementing an experimental plan
★ obtaining reliable data
★ analysing, interpreting and evaluating these data.

The second part of the examination is taken at a time specified by the examination board and is more open ended than the earlier exercise. Again, it consists of one or two questions.

This time you will be given apparatus with which you can perform rough initial tests for an experiment. From these tests you will develop a hypothesis and plan a detailed method to test it. This test is synoptic because you will have to draw on your knowledge of physics to support the experiment you carry out. Notice that you do not have to analyse these data.

You will also be given a description of an experimental method and a set of data (these may or may not relate to the experiment you have just devised). You will be asked to evaluate this described method, perhaps by making estimates of experimental uncertainties and so on. You will also be asked to discuss how the described method could be modified to give better reliability. Finally, you will be asked to analyse the data.

Edexcel (A)

Units with synoptic assessment	Title of unit	Percentage of *whole* A-level qualification
Unit 6 Paper time: 120 minutes	Synoptic test	20%

Table 2.3

The Unit covers the whole of the AS and A2 material with an emphasis on skills and principles rather than detailed knowledge and content.

Passage analysis is about 37.5% of paper.

Long structured questions are about 62.5% of paper.

Unit 6 contains two topics: Analogies in physics (springs and capacitors; electric and gravitational fields; capacitor discharge and radioactive decay) and nuclear accelerators (mass-energy conservation; linear accelerators; ring accelerators; nuclear particle detection).

Three questions are set. The first question will be based on the material in the topics. The remaining questions will draw on other areas of the specification.

Edexcel (B) – Salters Horners Physics

Units with synoptic assessment	Title of unit	Percentage of *whole* A-level qualification
Unit PSA5I	Project	5% (20 marks out of the total of 40 marks)
Unit PSA6 Paper time: 90 minutes	Synoptic test	15%

Table 2.4

Unit PSA5I

The parts of your project that are assigned to synoptic assessment are:

A Research and rationale   8 marks
C Implementing   6 marks
E Interpreting and evaluating   6 marks

Unit PSA6

This is a comprehension and data analysis test that covers the entire AS and A2 specification. You will be required to bring together knowledge and understanding from different areas of physics.

There will be a set of questions relating to a passage taken or adapted from a scientific book or article.

Additionally, there will be a compulsory unstructured question and a structured question.

OCR (A)

Units with synoptic assessment	Title of unit	Percentage of *whole* A-level qualification
Unit E Paper time: 90 minutes	Option paper, one paper from Units E1–E5	5%
Unit F part 1 Paper time: 60 minutes	Synoptic paper	10%
Unit F part 2 Paper time: 90 minutes	Experimental skills	5%

Table 2.5

Unit E

The options are:

E1 Cosmology
E2 Health physics
E3 Materials
E4 Nuclear and particle physics
E5 Telecommunications.

The question papers have the same format as those that you took for AS. They contain compulsory questions with both structured parts to the questions and parts that demand extended writing. The balance of marks is roughly 75 marks for the structured sections and 15 marks for the extended writing.

Two questions are set that include comprehension and data analysis. About 30 marks (one-third of the marks) are allocated to synoptic questions.

Unit F1

All questions demand extended writing. The questions are compulsory and test synoptic skills.

Unit F2

Your experimental skills will be assessed

either through coursework (marked in your school and college and moderated externally)

or through a practical skills test which is set and marked by the awarding body.

Practical skills test

There are two parts to the practical skills test: a Planning task and the main Practical itself. In the Planning phase you will be given a task; this will be set some time in advance of the main group of examination papers (exact arrangements are decided by your examination centre). You may be given laboratory time in order to try out some of your planning ideas. You need to write no more than 1000 words for this task.

The practical examination itself involves an experiment similar to the task you were asked to plan, but it will not be the actual experiment you planned. You will receive marks for the way in which you take observations, for your analysis of these observations and, finally, for your evaluation of the experiment.

OCR (B) – Advancing Physics

Units with synoptic assessment	Title of unit	Percentage of *whole* A-level qualification
Research report (part of Unit 5: Field and particle pictures)	Research report	7.5% (only ⅔ of the paper is counted towards the synoptic assessment
Unit 6 [paper 2865] Paper time: 90 minutes	Advances in Physics	15% (80 marks)

Table 2.6

Unit 6

Unit 6 [paper 2865] is a synoptic paper that gives you the opportunity to show that you can draw together ideas in physics and link different aspects of the subject. The paper includes a comprehension/data-analysis exercise.

Section A of this paper includes a printed text about physics (about 1500–2000 words) which includes data and diagrams. This will be given to you some weeks before the examination so that you have the opportunity to study the passage in advance. In the examination you will receive a new copy of the passage together with the (unseen) questions about the content. This form of comprehension/data-analysis paper allows you to have plenty of time to read the passage and look up unfamiliar ideas before sitting down to answer the questions. Whilst the context of the passage might be unfamiliar, the physics in it will be drawn from across the whole A-level course.

Section B of this paper contains two structured questions. These test synoptic abilities further through questions about physics and its applications in the context of problems that bring together ideas of physics from different areas of the course.

WJEC

Units with synoptic assessment	Title of unit	Percentage of *whole* A-level qualification
PH5 Paper time: 90 minutes		15%
PH6 Paper time: 120 minutes	Synoptic paper	12.5%
PH6	Investigatory task	7.5%

Table 2.7

PH5

This paper tests the topics of fields of force, electromagnetic induction, radioactivity, nuclear energy and probing matter.

The test has five short answer questions testing only Unit 5. There are two longer questions that are based on Unit 5 but which draw on knowledge and skills acquired throughout the whole course.

PH6 – synoptic paper

Structured questions are set that are designed to examine your knowledge of the whole specification.

Questions 1 and 2 (10 and 15 marks, respectively) will be structured questions to test your knowledge and skills from the whole A-level course, including reference to physics principles from any part of the course and relationships between the teaching units.

Questions 3 and 4 (20 and 30 marks, respectively) will involve questions based on a short passage relating to applied science and also questions involving data analysis.

PH6 – investigatory task (45 marks)

This is one piece of coursework of an investigatory nature, set internally by your centre but assessed externally.

Summary

Comprehension test or passage analysis possibly with some elements of data analysis:

★ All specifications *except* AQA (A).

Structured questions across the whole of the core specification:

★ All specifications.

Experimental skills:

★ AQA (B) (practical test; 5% of whole examination)
★ Edexcel (B) (part of project work; 5% of whole examination)
★ OCR (A) (either coursework or practical test; 5% of whole examination)
★ WJEC.

Written research report:

★ OCR (B) (project work; 5% of whole examination)
★ WJEC.

CHAPTER THREE

EFFECTIVE STUDYING
AND REVISION

During the synoptic A2 examinations you will need to recall and use information that you learnt early in the course. You need to acquire good quality learning skills from the beginning of A-level. This chapter discusses the skills of studying and revision as they apply to physics. These skills are not related solely to synoptic skills, they can be used in all areas of your learning as well as in other subjects that you study apart from physics. What *is* important is that you should think carefully about the way that you study. You should, if possible, start this thinking process early on. Good effective learning from the early Units of AS onwards will ensure your success not only in the synoptic parts of the examination but in the earlier parts too.

Helping your brain learn

The brain is extraordinary. It has about 10^{12} nerve cells (neurones), each with the opportunity to connect to the others. The total number of connections may be as large as 10^{800} – a number far bigger than anyone can appreciate. Your job in studying is to allow these connections – memories – to form and to stay formed. This, as far as the brain is concerned, is a chemical issue.

This is not the place for a detailed account of how the brain works – you can find that elsewhere if you are interested – but you should know that there are two sorts of memory: *short-term memory*, which is essentially an electrical pattern not yet 'frozen' in place, and *long-term memory*, which is a chemical imprint in the brain formed as a result of the short-term electrical discharges. You need to have a way of turning the short-term ideas you collect lesson-by-lesson into the long-term understanding of concepts and ideas you will need for examinations.

Memory and learning

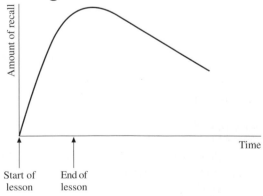

Figure 3.1 Memory recall with no review

Look at the graph (Figure 3.1). It probably tells you what you already know. A short time after a lesson or a piece of learning, your recall of the information is good and improving. Unfortunately, left alone the memories do not continue to improve in this way. After reaching a maximum, the level of recall starts to drop and (unless you take steps to prevent it) you will eventually forget the information you have learnt. The material has not moved from short-term to long-term memory.

Now look at the next graph (Figure 3.2) which shows what happens if you review material after a short break.

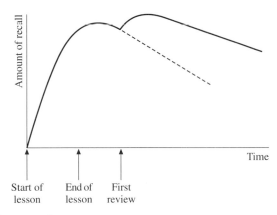

Figure 3.2 Recall with one review

This time the level of recall goes higher and persists for longer leaving a higher level of understanding for longer times. The third graph (Figure 3.3) shows what happens if this review is repeated time after time.

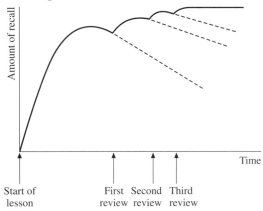

Figure 3.3 Recall with many reviews

Carrying out a review of your notes and your understanding at regular intervals has a very significant effect on the amount of long-term memory that is created in your brain. You should try to build review into your everyday schedule of work. To do this you need to organise when and how you study.

When to study

There is no right or wrong answer to the question of when you should study. It really is a matter of personal preference and, just as important, the time of day that suits you and your brain best. Some people work best in the morning, some at night. Students are often stereotyped as people who learn during the day and do homework at night. This may not be right for you, though. Some people are more effective before the evening meal, some after it. You should try different options and come to a decision about the best and most productive time of day for your study. You may well find that your best study time is first thing in the morning. If so, get up early and do your work then.

How to study

As well as time of day, it is important to get the physical conditions right in order to allow good learning to occur. Have you a place where you can store your materials and files, where there is enough light, and where you are comfortable? A cluttered desk where you have to search hard to find a pencil or ruler is probably not the best work area for study.

Make sure that you have everything to hand before you begin – you have to know what you intend to study before you start. Obvious, but many students are haphazard and unprofessional about the way that they approach what is actually their full-time job. You need a timetable and a plan of action.

Study plans

Buy or make a day-by-day calendar that runs until the end of your A-level course. Find out when you are likely to have major internal tests at school or college, and also when you will be taking the external Unit tests for the course. Put all these dates on the calendar. Work backwards to establish when you should begin revision for these tests. Pencil in what topics will be examined. This is all-important planning for revision.

Now include details of the reviews you need to carry out. Some of this will be routine. Every day you will want to review the day's lessons and follow these up with more study or with the assignments set by your teacher.

There will probably be additional pieces of work that will take an extended length of time: coursework, essays to write, past papers to prepare. As these are assigned to you, put the completion dates onto your calendar and indicate the range of time over which you intend to carry these out. Allow enough time and try to break the task down into manageable chunks as far as possible.

This study planner is for medium- to long-term strategic study planning. You also need to operate on a day-by-day basis to keep on top of your work and study. Be realistic about how much you can actually achieve in the two or three hours of private study time you carry out each day and decide when you should schedule this time. The small amount of time that you will devote to preparing and updating a study planner will be more than paid off in increased effectiveness in your study and learning.

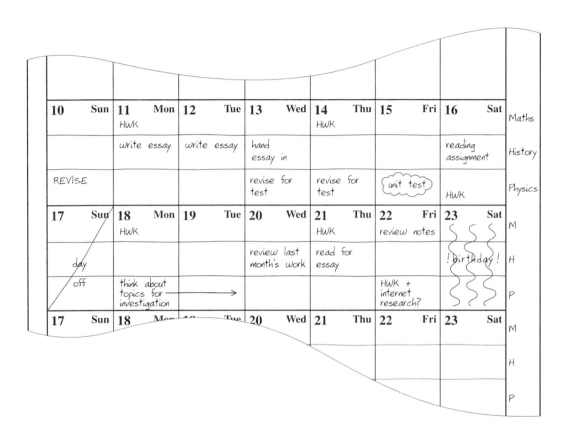

Figure 3.4 Part of a student's planner

Give yourself a break

Here are some pointers to the effective use of a chunk of study time:

★ During the first few minutes of a study period, the brain settles down to work and this time should not be devoted to important learning or writing. Use the first few minutes for less important study, say review of all the day's lessons.

★ Take breaks in your work. You should definitely plan to leave your desk and take a break of no more than a few minutes at regular intervals. To some extent, the time between breaks depends on the work you are doing. A repetitive task, graph drawing or the preparation of diagrams for a report, say, could have longer times between breaks than difficult learning or substantial calculations. As a rule of thumb, aim for 20 minutes for difficult work and between 40–60 minutes for repetitive work.

★ During the break, consider moving to a routine topic or job (filing notes), moving around and away from your desk, relaxing for a short time. Do not take too long a break as this will mean that you have to do some brain-settling again. Five minutes should suffice – enough time to make a cup of coffee, definitely not long enough for a telephone call to a friend!

★ Plan the most difficult work for the beginning/middle of the study period. Work at the beginning or end should be more mundane or routine.

Imagine that you have the following work to do. Prioritise it over two consecutive weekend study periods of 3 hours each. Aim within each study period to have four study chunks with three short breaks of 5 minutes each.

Work	Carry out equation of motion calculations	Review week's work	Revise some topics for Unit test	Write up experiment for coursework portfolio	Start planning for major essay to begin after Unit test
Probable time to complete	90 min	60 min each week	120 min	240 minutes	30–60 minutes
Due in	After first weekend		Test in one month	In one week	

Table 3.1

Hint There is no right or wrong answer to this question. But it does illustrate the amount of juggling you need to carry out to satisfy all the demands of an A-level course and the demands of all your different subjects. There is about 11 hours work to be carried out in the 6 hours available so you need to be ruthless. The remainder will have to be carried out during the working week.

As a rough guide:

★ Schedule the week review into four short chunks at the beginning and end of each 3-hour study period – the review is important but not high level.

★ The calculations are a priority: put them early in the first study session so that there is a safety net if they are hard or take longer than you think.

★ The write-up is important and must not be left for too long, but you could begin work on a first draft and aim to polish it up during the following week. Equally, diagrams and graph drawing can be left as mundane tasks for next week. Perhaps 120 minutes at this stage, spread across two study sessions is about right.

★ The revision for the Unit test is important. Timetable it late into the first session, but early into the second. This revision should be the subject of its own schedule.

Look after yourself

'All work and no play makes Jack a dull boy.' And this applies no less to Jill. It is important that you take time for recreation and that you reward yourself for the study that you undertake.

Taking time out

The body and the brain need oxygen and exercise. The brain takes about 25% of the intake of oxygen for the whole body. If you are an active sports player, keep your sport going. Make time for it in your schedule and regard it as an important part of your work. If you do not play a sport at the moment, consider doing so. Perhaps take up cycling or walking to and from school or college.

This need for recreation is especially important around the time of external examinations. Examinations stress everyone and this stress leads to the production of adrenalin and its by-products. By far the best way to remove these chemicals from the body is by taking exercise, so do not stop taking exercise during the period of the examinations. If anything take more!

Reward yourself

Try not to make study an unpleasant chore. Agreed, it is hard, demanding work and sometimes there are other things you would rather be doing, but use this to your advantage. Any prolonged period of study deserves some reward. The choice of reward is up to you, whether a phone call to a friend or a visit to the cinema probably does not matter. Whatever you decide, though, do not let your hard study go unrewarded, if you think you have earned a reward, then take it.

Effective learning: some key points

- **Plan, plan, plan – realistically.**
- **Keep a day-by-day chart and stick to it – but allow flexibility in an emergency.**
- **Take breaks when you study.**
- **Use what you know about how the brain works to enhance your learning.**
- **Do difficult, high priority work near the beginning of a study session.**
- **Do low-grade repetitive work near the end of a study session.**
- **Keep active – take time out for recreation.**
- **Reward yourself for study.**

Written study skills – making (and keeping) good notes

An important key to success in developing your synoptic skills is good access to information that you have already studied. The full A2 course lasts almost two years and at the end you need to recover notes that you made at the start of the course. Again, organisation is the key.

At GCSE your teachers will have controlled the type and the quality of your notes very closely. Although they will continue to control your work as you move through the A-level course, they will expect you to manage your own note-taking more and more. You will also need to develop your own note-taking style. These are all parts of the process of becoming an effective A-level physicist.

Your own tastes will determine how you record your lessons. The following are some examples of the types of notes that you could generate.

Linear notes

These might be the types of notes that you take during a lesson and most people would think of these as traditional notes. They are likely to follow the order in which material is presented to you.

However, they are not necessarily the most appropriate notes for long-term revision purposes. If you intend to re-work notes either soon after taking them or later for revision you might consider presenting the material in a more visual way. Some of these ways are shown later.

Another possible idea to allow incorporation of extra notes later is to write only on one side of the page in your notebook or on one side of the sheet if you use a ring binder. (An alternative is to divide each page into two halves with a vertical line – then make the original notes on one half of the page keeping the blank half for later additions.) This will allow you to supplement your original notes with additional material later. These additions could be examples of calculations that you perform later or extra material you derive from textbooks during later study.

Sprays

Sprays are useful when you are brainstorming and thinking through a complex problem. They may be helpful, for example, if you have a research report to write and you are beginning to think through the topics and the physics that you need to consider further. Begin by writing down – in no particular order – the ideas as they come into your head. Then, perhaps after a few hours' reflection, you can join up the separate ideas to give a coherent thread to your report (Figure 3.5).

Step 1. Getting ideas down

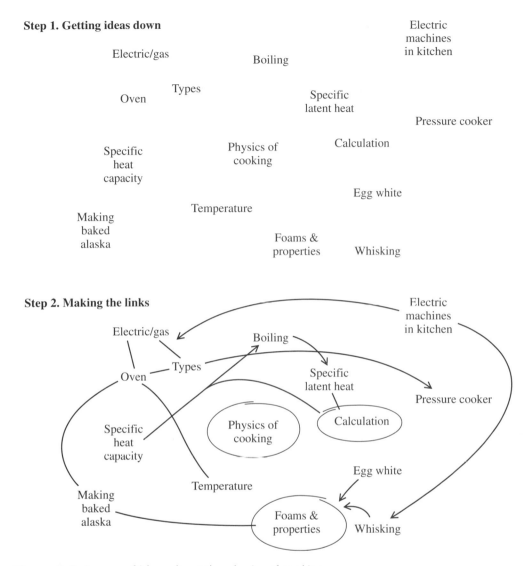

Figure 3.5 A spray of ideas about the physics of cooking

Visual notes

Pictures are always a good idea. They can be used to sum up many pages of linear notes and are a good way to organise your ideas. They can be based on a picture of a piece of physical apparatus, or perhaps on a graph, and can be annotated to include the relevant pieces of information pertinent to that physics (Figure 3.6).

Patterns (sometimes called spider diagrams or concept trees)

Patterns are a powerful way of organising information and can be used to organise both permanent notes and 'brainstorming' sessions when beginning a piece of work (Figure 3.7 shows the beginning of a pattern about the inverse-square law, you might want to complete it).

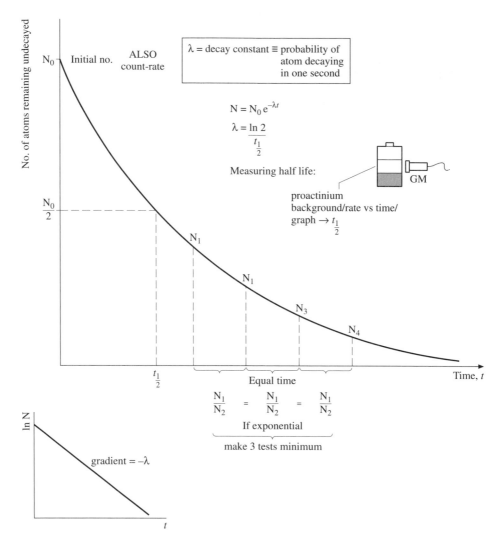

Figure 3.6 Radioactive decay curve – some visual notes

It has to be said that not everybody feels comfortable with notes made this way. But, as in all things, practice makes perfect. It is a technique that can be used – if only on a small scale – to brighten up your linear notes at appropriate moments.

Card notes

Some people like to transfer all their formal notes on to a set of 5″ × 4″ cards for revision purposes (Figure 3.8). This has the advantage that they can carry the cards about easily and unobtrusively, and can spend odd moments learning when the opportunity arises, on public transport for example.

The sheer bulk of your class notes can be very daunting during final revision. Transferring the notes from a linear to a card form will help you to order the information in your head and the small volume of the cards will give you the confidence that you can learn this material.

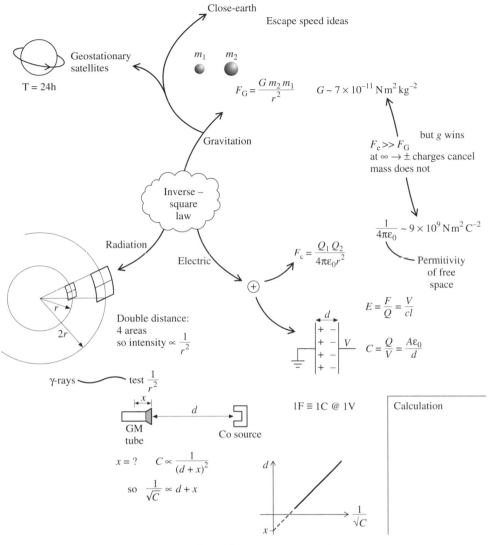

Close-earth

Escape speed ideas

Geostationary
satellites

$T = 24h$

m_1 m_2

$F_G = \dfrac{G\,m_2\,m_1}{r^2}$ $G \sim 7 \times 10^{-11}\,N\,m^2\,kg^{-2}$

Gravitation

but g wins

$F_e \gg F_G$
at $\infty \to \pm$ charges cancel
mass does not

Inverse –
square
law

$\dfrac{1}{4\pi\varepsilon_0} \sim 9 \times 10^9\,N\,m^2\,C^{-2}$

Radiation

Electric

$F_e = \dfrac{Q_1\,Q_2}{4\pi\varepsilon_0 r^2}$

Permitivity
of free
space

r

$2r$

Double distance:
4 areas
so intensity $\propto \dfrac{1}{r^2}$

$+$

$E = \dfrac{F}{Q} = \dfrac{V}{cl}$

d

V

$C = \dfrac{Q}{V} = \dfrac{A\varepsilon_0}{d}$

γ-rays   test $\dfrac{1}{r^2}$

x

GM
tube

d

Co source

1F ≡ 1C @ 1V

Calculation

$x = ?$ $C \propto \dfrac{1}{(d + x)^2}$

so $\dfrac{1}{\sqrt{C}} \propto d + x$

d

x

$\dfrac{1}{\sqrt{C}}$

Figure 3.7 The beginning of a pattern about the inverse-square law

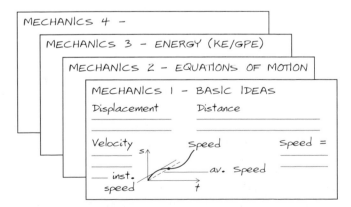

MECHANICS 4 –

MECHANICS 3 – ENERGY (KE/GPE)

MECHANICS 2 – EQUATIONS OF MOTION

MECHANICS 1 – BASIC IDEAS

Displacement   Distance

Velocity   Speed   speed =

inst.   av. speed
speed

Figure 3.8 Card notes on mechanics

Key points

Make notes visual. A picture is said to be worth a thousand words and, on a time basis alone, is much quicker to construct. With practice you will be able to include all the relevant information and more on a clear, labelled and well thought-out diagram.

Use colour in any diagrams that you draw for note-taking purposes. It helps enormously with your memory. The more that you can force your brain to operate in different modes the better.

Re-working notes helps you to realise what you do and do not know. It helps you to focus on the fundamentals.

Working on notes – re-writing, condensing, making pictures – is active learning; just reading notes through is not. Active learning is the best form of revision. There is more about this in the next section.

NOW TRY THIS

Here are some suggestions of things you might try to help you experiment with note taking. (Activities marked with a * are more suitable for someone at A2 level.)

1   Look at some linear notes you took some time ago. Have you been taught anything connected to this topic since? Have you added to the original material in any way? Have you added references to pages in your textbook where these topics are covered? Are there any ideas that you have found hard – do you need to do more reading around the topic or ask your teacher? Are there any calculations you did for homework that could usefully be copied into these notes?

2   Try making a spray on the following topic:

A small coastal village community in Scotland has to choose between a wind-powered and a tidal-powered energy generating station. Brainstorm the advantages and disadvantages of these two types of station for the village.

When you have finished, persuade a friend taking physics to carry out the same exercise and then compare sprays.

3* Try making a visual note of a direct current electric motor. Your diagram should show the motor itself with all its relevant parts (photocopy a diagram from a book if you find the motor hard to draw). Label the diagram. Add a visual indication of how the commutator works. Include the principle pieces of physics in as pictorial a way as possible.

4   Construct a pattern (concept tree) for the mechanics you have studied in physics so far (that is, up to GCSE for AS students and as far as you have got, for A2 students). Use colour; keep everything as visual as possible.

There are no formal answers to these activities. Put your answers away for a few days and then look at them again. Did you miss anything out? Which note form has helped you to retain the concepts and ideas best?

Time to revise

So the time has come. You are near the end of the course and for the past two years you have been studying carefully with good study skills and have kept detailed and clear notes of the separate topics you have been learning Unit by Unit. This section begins with some advice on good revision practice (which applies to all Units not just the synoptic ones) and then goes on to look at the types of synoptic questions that are set in A2 physics.

Good revision

How do you revise at the moment? Take a few moments to consider how you handle the task of preparing for a test or a Unit examination.

Answer these questions:

How active is your revision?

Do you re-work notes in some way? Do you find this effective?

Does your revision begin early enough?

Do you have access to past papers?

Do you complete past papers or just give up?

Do you complete the papers against the clock?

There are no right or wrong answers here. But there are some pointers to good practice:

Your revision should be as active as possible. Simply sitting in a chair leafing through your existing set of notes is not very effective at all, yet this is how many people approach revision. You must make your revision sessions rewarding and active. You must feel a sense of achievement at the end of the session and you must feel that you have made progress in your understanding of the work.

Too many students simply read through notes and assume that this is a good way to re-learn material. In fact, it is a very inefficient learning method for physics. To score a high mark in a Unit, you will need to demonstrate an understanding of the material tested in the examination. An approach to revision that emphasises rote learning and pays little account of topic understanding is unlikely to yield top marks.

A better approach is to remember that the more you can involve your whole brain in learning, the better. Do not just *read* the notes and books. Instead, use one or more of the following strategies to revise the work:

★ Write a new set of notes, possibly in a different format from the original set.

★ Consolidate the notes into a spider diagram or into a card format.

★ Add material to your original notes (this assumes that you left space for additions when you originally wrote them).

★ Add to your notes page references to your favourite textbooks – do not force yourself to use the index in the book every time you refer to a topic.

★ Work through examples in a book or try past papers and then add these to your original notes.

Start early and pace yourself. Do not begin your revision too late. You need time to allow yourself to come to terms with new, and old, ideas. Some candidates sit Units in January whilst still learning new material. Your teachers will expect you to be able to juggle the demands of revising old physics and learning new topics. You cannot do this if you have not allowed enough time. Synoptic papers may be amongst the final papers you take; they use physics from the whole course. Remember that you will have been taught some of this material almost two years earlier. Give yourself time to come to terms with this material again.

Active revision means

● **Re-working notes**

● **Answering past papers (possibly against the clock as you get near to the examination)**

● **Pulling ideas together using new and different ways of presenting the material.**

Strategies for answering past examination questions

Teachers ask you to answer past exam questions for many reasons including:

★ to test that you have understood the work correctly

★ to test your overall standard at the end of a section of the specification or as you approach a final examination

★ to give you practice in answering real examination questions, possibly against the clock.

If you are answering questions set by a teacher, as opposed to questions you are answering as part of your own revision programme, make sure that you obey the teacher's instructions. The advice in this book may conflict with what the teacher wants you to do – but your teacher may well have other good reasons for setting the questions.

In answering past or specimen questions for your own revision try the following strategy:

TIME – READ – REHEARSE – WRITE – REVIEW – FILE

Time

If at all possible, find out how long the question is supposed to take. It may be that the question has a time allocated to it on the paper itself, in which case your job is easy.

It is sometimes more difficult to assess the time in the case of a paper that has several questions and just a mark allocation for each question.

Step 1. Find the total number of marks on the paper.
Step 2. Find the number of marks for the question you are attempting.
Step 3. Calculate the time available.

EXAMPLE

Timing an exam

Example 1

Suppose your examination paper lasts 90 minutes and contains 60 marks.

So, each mark is worth about 1½ minutes.

A 3-mark calculation should take 4½ minutes, call it 4 minutes with a 30 second check.

A 7-mark description of an experiment should take 10½ minutes, call it 10 minutes with a 30 second check.

Example 2

Section A

You are advised to spend **one hour** on this section

Answer both questions which have the same number of marks allocated to them

In this case, one question takes half-an-hour. Do not spend a minute longer on it!

Example 3

One of your papers takes 1¼ hours and has a total of 60 marks.

Question 3 has 11 marks allocated to it.

So question 3 should take *about* 9 minutes.

It is dangerous to run out of time in an examination. Answering seven out of eight equally weighted answers in a paper limits your maximum mark to 87.5%. Better to answer 87.5% of each of the eight questions. Why?

Exam questions tend to become harder the further into them you get. So you will probably score the easiest marks from all eight instead of completely losing the marks for the final question.

If you are just beginning your revision, do not worry too much if questions seem very difficult and take too long. First of all, practice makes perfect and you *will* speed up as time goes on. Secondly, everyone works faster under examination conditions when the pressure is on. So, taking a little longer, even quite close to the examination, is not necessarily a problem.

Read

Read through the question once.

Read it again (get used to this routine of two readings – it is designed to ensure that you really understand the demands of the question).

By this time, you should have realised whether you already remember enough to be able to answer the question.

If you can answer the question without recourse to any books or notes, do so. In other words skip the *Rehearse* step.

But, if you do not feel that you can answer the question, make a brief note of what you need to look up, and then put the question itself away.

Rehearse

Certainly in the early days of revision, you may not be able to answer whole questions, especially synoptic ones, without looking up all or part of the physics. Do not regard this as a failure. You are simply identifying areas of physics that, at this stage, need more revision work.

Use notes or books or anything else you find useful in order to remind yourself of the physics. These 'other materials' could include similar questions from books or questions you have done successfully in the past, they could also be CD-ROMs or access to reliable sources on the Internet. Work through these until you are happy that you understand the physical ideas contained in the question and you are confident that you can tackle the question.

During this stage you may be adding annotations to your notes or adding the past questions and worked examples to your work. Try to add enough material to your notes so that you will not need to go through this process again.

Now put your notes away.

Write

Answer the question – as far as possible against the clock. If there are spaces given on the paper for your answers, try hard to keep your written answer within these limits.

Review

When the answer is complete, review it to ensure that it has satisfied the demands of the examiner. Have you provided enough points for the examiner to mark? In other words, a three-mark part to a question will generally require you to give three points or three features in your answer.

File

Add your answer, together with the original question, to your file preferably near to the section of work from which it comes. You do not have time when preparing for exams to re-visit work twice!

Tips from the examiners (past paper reports and mark schemes)

Past-paper reports

Every year the examination boards issue extensive reports on the A-level examinations. These reports discuss the papers and the performance of candidates in them and are written by the examiners responsible for the individual papers. The reports contain detailed information about the topics on which students scored poorly, or mention common misconceptions revealed in the answers. Try to obtain these reports if you can, or ask your teacher to give you a summary of recent reports.

Reports can often give a clue to the examiners' expectations such as:

★ number of significant figures allowed in answers
★ penalties deducted for errors in units attached to numerical answers
★ standards required in graph drawing.

You will find advice about these topics later in the book.

Mark schemes

The boards also issue detailed mark schemes. These are the rules that tell the individual examiners how to award marks. Your teachers will, of course, not wish to give you these schemes at an early stage in your course because it will then be too easy for you to produce perfect answers in a test based on these past papers. But, late in your revision programme, you will find it helpful to look at these schemes (especially after you have answered questions and had the marked questions returned). As with the past-paper reports, you will be able to see why the examiners deduct marks and where these deductions are likely to occur.

The mark schemes should also convince you that there is one mark for each point in the answer. You should always read the mark allocations printed on the examination paper itself. If there are three marks assigned to part of a question, then there are three points for you to make. Get into the habit of thinking through the important points in an answer before you commit your answer to paper.

Learning and revising for synopsis

So how does all this fit together, to help not just the first Unit tests, but the all-important synoptic papers that you take at the end of the two-year course?

Synoptic Skills in Advanced Physics

Learning for synopsis

If you are going to score well on a synoptic paper you must have a good understanding of the key ideas and concepts used in A-level physics. Learn not just the facts about your physics but also work at understanding the principles that underpin the subject. This fundamental understanding does not come naturally, you have to encourage it. Get into the habit of making these connections throughout the A2 course as you build up a larger and larger repertoire of topics.

Here are some ways to help you to reinforce these underlying concepts:

★ Incorporate synoptic notes into your existing folders and files or, if you prefer, have a separate set of notes that deal specifically with synoptic topics.

★ Think carefully about the 'big' ideas in physics that might provide you with a starting point for these concept notes.

'Big' ideas in Physics

Here is a short list to get you thinking:

● **Units**

● **Vectors**

● **Fields**

● **Exponential changes**

● **Energy**

● **Momentum**

● **Graphs**

● **Oscillators.**

There are many others that you may prefer to these – you can probably think of some straight away. The important point is for the notes to suit you and to help you consolidate all your learning experiences throughout the two years of your physics course.

★ Use sprays/spider diagrams to build up visual connections between topics.

★ Revise the synoptic notes regularly. A good time for this is just after you have completed a topic in class.

★ Get used to thinking in ideas and concepts not just separate and distinct topics.

Potential difference (voltage) is an idea used in electricity theory.

But it is really an extension of electrical ideas into more fundamental ideas about energy. How does electrical potential difference link up with the ideas of:

★ Energy

★ Gravitational potential energy

★ Graviational potential difference?

In what ways do the units of these all these quantities differ?

In what ways are the units similar?

Finally, Figure 3.9 is an example of some incomplete concept notes (in the form of a spider diagram) made by a student who was consolidating ideas about exponential changes. You might like to copy and finish this diagram.

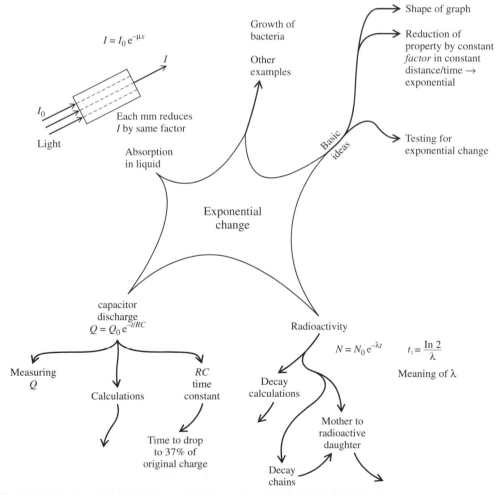

Figure 3.9 Exponential changes – an incomplete spider diagram

Revising for synopsis

★ Think broadly. In other words, keep synoptic skills in your mind as you revise. Try to think about the synoptic implications and possibilities of a topic whenever you are revising a specific topic in depth. Ask yourself how could this topic be combined with others to produce a synoptic question?

How could an examiner extend a question about the following topics to test your synoptic skills?

★ Discharge of a capacitor

★ Alpha-particle scattering

★ Theory of a cyclotron

★ Wave motion along a slinky string.

★ Know your specification. When and how will synopsis be tested? What will the shape of the questions be?

★ Mix topics as you revise. Try not to work sequentially through your notes, but move around and try to combine material from different years of the course.

★ Revise by concept, perhaps not every day but certainly once or twice a week.

Revising by concept

Here is an approach you might experiment with:

Choose a 'big' idea from the whole of physics. On a blank sheet of paper take no more than 2–3 minutes to write down – in no particular order – all the facts and ideas you can think of that relate to this 'big' idea. If you have a set of notes in a reasonably compact form, lay out all the notes relating to these various parts of the subject. Read quickly through them.

Now to make the revision more active:

Try to write a question that brings in as many ideas as possible.

Find some questions that are either synoptic or very broadly based and answer them.

Remember, the most effective way to revise – whether for synopsis or anything else – is to make your brain *use* the material.

★ Finally, answer plenty of synoptic questions under test conditions.

Remember the earlier advice

● **Do not expect to be up to speed straight away.**

● **Do expect to have to read the question carefully, understand its demands, look up the material and then write the answer. At least in the early days.**

● **Do look at the specimen mark schemes to realise the level of answer that you are expected to attain.**

NOW TRY THIS

Here are just a few synoptic topics you could explore for yourself during your revision. There are no formal answers. The first few topics indicate some possible areas of study:

★ TV sets – e/m, flow of charge, motion of charges in magnetic/electric fields, mechanics …

★ Swimming pools – estimates of volume, buoyancy ideas, pressure, specific thermal capacity ideas …

★ Boats – mechanics, drag, mechanical/electrical power of motors, satellite navigation …

★ Gieger-Muller tubes

★ Motor cars

★ Energy changes and resources

★ Circular motion

★ Material properties of all kinds

★ Inverse-square laws

★ Exponential changes.

CHAPTER FOUR

CONNECTING USING WORDS

This chapter looks at the skills required in writing examination papers and in producing research reports. Other skills that you need (mathematical, graphical and analytical) are discussed in Chapters 5 and 6. The additional skills needed for practical examination papers and practical investigations are discussed in Chapter 7. You might need these later skills whilst you are answering some of the questions and considering the points mentioned in this chapter; if so, you should look ahead.

One further point: much of the advice in this chapter goes beyond the assessment of synoptic skills. You will find that the advice here is relevant to your work in other parts of your physics, for example, the section on trigger words is relevant to the non-synoptic examinations in both AS and A2.

Trigger words in questions

Read the question carefully. How many times have you been told that? Examination questions contain so-called 'trigger' words or expressions that flag up what the examiner actually wants. It is important to understand as precisely as possible what is required. Here is a list of some of the trigger words – always remember, however, that there are alternatives (some listed here) and examiners will use a different trigger word if they feel that it makes a question clearer.

(Some typical questions below have answers in note form; you should, however, normally aim to write complete sentences in an examination)

Calculate *or* **Evaluate** *or* **Determine** *means*

Carry out a calculation to derive a numerical answer.

Question: Calculate the charge that flows through a resistor in 5.0 s when the electric current in it is 2.5 A.

Answer: $Q = It = 2.5 \times 5.0 = 12.5$ C

Compare *or* **contrast** *means*

Put both sides of a case.

Question: Compare the advantages and disadvantages of solar cells as the energy source for an island off the west coast of Scotland.

Define *means*

Supply a definition of the term quoted.

Question: Define the volt.

Answer: A potential difference of 1 V exists between two points if 1 J of work is done when 1 C of charge flows between them.

Describe *means*

Put into words what is meant by a physical effect or an experiment or the shape of a graph.

Question: Describe the motion of a motor car when a constant force acts on it.

Answer: There is constant acceleration so speed increases steadily.

Design (*usually in the context of an experiment*) *means*

Devise and then describe an experimental procedure by which you can determine some physical value or verify a relationship.

This type of question will often go on to specify the information that is required by the examiner – read this very carefully and ensure that you do provide exactly this information.

As a general rule, an experimental design will require:

★ a **clear diagram** of the proposed apparatus

★ a **description of the method** you intend to use

★ a description of the **data** you intend to collect

★ a **statement of the way in which you intend to use these data** – this is often best shown as a sketch graph perhaps showing the graph trend you expect to observe.

If you are taking an examination in which you then go on to carry out the experiment, you may also be asked to **evaluate** the experiment.

Question: Design an experiment to show that, for small oscillations, the time period T of a simple pendulum is related to its length l by $T^2 \propto l$.

Discuss *means* …

… **that the examiners require an extended answer**. You will need to show knowledge and understanding of the topic in the question. You may be required to provide a balanced view giving both sides of an argument.

Question: A farm needs an electrical supply that is independent of the mains in the event of failure of the main supply. Discuss the advantages and disadvantages of wind power as a backup supply for the farm.

Answer: The answer would need to include:

Advantages – e.g. cheap once installed, reasonable set-up cost, low maintenance, etc.
Disadvantages – seasonal, no supply at night, can depend on cloud cover, would need additional storage (e.g. storage cells) to cover all eventualities, etc.

Estimate the size of … *or* **determine the magnitude** of … *means*

You are asked to work out an answer that is only approximate. In the course of the calculation you may have to provide an educated guess of some quantities that are not given in the question.

Always make clear what values you have needed to guess in this way.

Question: Estimate the force exerted on soft ground by a man jumping to the ground from a car roof.

Answer: There are a number of ways to answer this question, here is one of them:

Step 1 make estimates: mass of man 80 kg
height of car roof 1.5 m
dent made in ground = 0.1 m deep

Step 2: calculate energy lost = *mgh* = 80 × 10 × 1.5 = 1200 J

Step 3 *work done = force × distance* 1200 = *F* × 0.1; *F* = 12 kN

Explain *or* **justify** *means*

Give a reason usually after making a statement. These trigger words are often coupled with 'Describe …' or 'State and …' In this case *always* begin your answer with a clear statement/description of the physical point.

Question: State and explain what happens to the resistance of a metal as its temperature increases.

Answer: Resistance increases. Atoms vibrate more as temperature rises. Charge carriers (electrons) are increasingly impeded by this vibrational motion. Current drops for given potential difference and so resistance increases.

Question: A student suggests that the electrical resistance of a thermistor drops as its temperature increases. State whether you agree with this statement and justify your answer.

Illustrate *means*

In the context of a written question, **give some examples of the physical point you are discussing.**

Question: Explain what is meant by a *renewable resource*. Illustrate your answer with three examples of renewable resources.

Answer: An energy resource that can be regenerated in a short time. Examples include biomass, tidal energy sources, wind-powered energy sources.

List *means*

Give short answers in a note form (possibly in a table).

Question: Give a list of the SI prefixes that correspond to the following powers of ten: 10^{-3}, 10^{-6}, 10^{-9}.

Answer: m for milli; μ for micro; n for nano.

Name *or* **Give** *means*

Write a very short answer of just a few words in which you are asked for a technical term.

Question: Name the unit of radioactivity.

Answer: Becquerel.

Outline *means*

Give a brief account, possibly in note form.

Question: Outline the steps involved in measuring the time period of a simple pendulum.

Answer: Decide on fiducial mark (centre of swing); start oscillations; start clock as pendulum passes fiducial point; stop after 10/20 oscillations; repeat measurement; take average.

Plot *means*

Draw an accurate graph using data points that may be given to you or that you may have calculated.

Show that ... together with an approximate numerical answer *means*

Give working that arrives at the answer quoted in the question.

Question: Show that an object falling for 2.5 s with an acceleration of 9.8 ms^{-2} reaches a speed of about 25 ms^{-1}.

Show that ... together with an algebraic expression *means*

Prove algebraically the answer quoted in the question.

Question: A body of mass m starts from rest and gains a speed v. Show that the kinetic energy gained by the body is $\frac{1}{2} mv^2$.

Sketch ... a diagram *means*

Draw a diagram and label it fully. At A-level you may not be specifically asked for labelling. Don't forget to do so.

Question: Draw a circuit diagram of a circuit that would allow you to plot a graph of the potential difference across a resistor against the current in the circuit.

Sketch ... a graph *means*

Draw a diagram showing the shape of the graph on correctly labelled axes. Include any data points from the question or from calculations that you have carried out.

Question: Sketch a graph of potential difference against current for a metal wire.

State *means*

Give a brief description of the main point of the question. Often followed by '... and explain'.

Question: State the principle of moments.

Question: State and explain what happens during the discharge of a parallel-plate capacitor.

Suggest *means*

Give an outline proposal or explanation, nothing detailed is required.

Question: Suggest why it is impossible for a geostationary satellite to be put into orbit over Bristol.

Answer: The gravitational force vector is towards the centre of Earth not in line with the centripetal vector which needs to be towards the centre of the proposed circle.

Tackling writing in examinations

Writing your answers

It is common to find candidates putting themselves at a disadvantage through presenting their answers poorly – and this is not just a question of producing neat work in legible handwriting. As an example, examiners may provide you with lines for your answers in the form of blank lined answer books with separate examination question papers. Alternatively, you may be given a combined question and answer book where the lines for written answers are printed immediately below the question itself. There are merits to both types, indeed you may receive a hybrid paper in which there are substantial numbers of lined pages provided for extended essays – perhaps at the end of the question booklet.

Lines interleaved between questions

Examination printers are usually generous with the provision of lines for answers and there will be adequate space even for candidates with very large writing. The moral is: if your writing is small, do not feel obliged to fill all the space. What is important is to develop the skills of writing concisely and to include all the points required.

If you do find yourself regularly running out of space in mock examinations and tests, ask yourself why. Is it because your handwriting really is large or, alternatively, are you spending too much time on questions and failing to complete the later questions in the examination? Chapter 3 gives advice about answering timed questions.

Lined answer books separate from the question paper.

Here you need to be more careful. How many lines can you write in 5 minutes including thinking and preparation time? If you do not know, try to find out under test conditions. Remember the answer and use this knowledge when you answer questions on this type of stationery.

With lined answer books and no indication of a maximum length for an answer it is even more important to keep your eye on the clock.

But whichever stationery you use keep your answers neat and in an obvious place. If you do run out of space on the question paper, make it clear to the examiner where the remainder of the answer is. Although the examiner will make every effort to give you marks, if your work is simply not clear then you cannot expect to gain credit. If you *must* write different parts of a question on different pages of answer books (or even in different answer books!) make it absolutely clear where the answers are. Number the pages if not already done so and make life easy for the examiner.

If you are asked to fill in boxes indicating the order in which you answered questions, do so – it only takes a moment after the examination ends. If you are asked to rule a line at the end of a question, again try to do it – it is a courtesy to the examiner and takes only a moment of your time. Try to keep the examiner on your side as much as possible by writing neatly and making all corrections clear. It may be worth a mark or two to you if work remains legible.

Spelling, punctuation and grammar (spg)

There is a requirement for your spelling, punctuation and grammar to be examined. The way in which this happens varies from specification to specification. Your teachers will make it very clear to you how this will be carried out. Often, one or more questions on the examination paper will be designated as targets for spg and, as well as marks for the physics, you will receive marks for your use of English. This is often flagged up in the question itself using a form of words like:

'Two of the seven marks in this question are for the quality of your communication.'

It is worth reading through such questions when you have finished them, looking for errors in spelling and punctuation. Try, whilst you are writing, to ensure that you give a logical answer with no repetitions or loose usage of English or physics. Examiners will usually be generous (they realise that examination candidates write against time and under pressure) but they will not be allowed to give credit for ill-written or negligent pieces of prose.

Marks and timing

Most modern examination papers give students information about the number of marks available for a particular question. Know where this information is (often in the right-hand margin) and use it to plan both your answers and the length of time you spend writing them. Some boards give detailed information sub-section by sub-section, whilst others are less helpful and only give information about a whole question or part of a question. Whatever level of information you get in your examination use it as intelligently as you can in order to:

★ assess how many points of physics are needed
★ plan your time
★ plan how much you should write.

Structured questions with short pieces of writing

If you know the physics well, these are usually the easiest questions to answer.

★ Look at the mark allocation, this gives you the time and the number of mark points available.

★ Focus on the trigger word and ensure that you really are answering the question that has been set.

★ Rehearse in your mind what these important physical points are. Briefly put them in logical order.

★ Write concisely and to the point.

★ Check quickly what you have written and move on.

Extended writing

You may well be expected to produce at least one piece of extended writing in your examinations. This can range from a 10-minute comparison of a few energy sources to a 30-minute essay giving the physics of, and a few examples of, exponential change. Whatever the length, you need to approach writing of this sort in a methodical and structured way.

First of all, establish what the examiners require. As usual this means reading the question carefully and looking for hints for structuring your answer.

It helps to have ready a standard essay outline that you can always use in an examination (with some adjustment from question to question). For example, this could be a five-paragraph format:

ESSAY PLAN TEMPLATE

Introduction:

New paragraph 1: Main idea 1

New paragraph 2: Main idea 2

New paragraph 3: Main idea 3

New paragraph 4: Main idea 4

Conclusion:

Remember that in good writing, each paragraph consists of one main idea usually found in the first sentence. Later sentences in the paragraph develop the theme and should follow each other in a logical order. The first sentence should either be preceded by a blank line or should be indented.

NOW TRY THIS

The passage below is not capitalised or paragraphed. First of all, notice how difficult it is to assimilate. Next, decide where you would put the paragraphs.

radioactivity is the random decay of atoms which change into new elements with the emission of other particles and electromagnetic radiation. there are three principal emissions: alpha and beta particles, and gamma radiation. alpha particles are helium nuclei, two protons and two neutrons ejected from the nucleus. the helium nucleus is a particularly stable entity and is ejected as such rather than as the individual nucleons. as a result of alpha emission, the proton number of the original radioactive element drops by 2 and the nucleon number falls by 4. Beta particles are electrons emitted from the nucleus. the electron appears when a nuclear neutron decays into a proton. the proton number here goes up by 1 and the nucleon number stays constant. gamma radiation is emitted as a result of the loss of energy by the new nucleus, normally created in an excited state. gamma radiation is a high-energy electromagnetic radiation with a short wavelength and high frequency. radioactive elements decay with a half-life. this is the time taken for half of the original atoms to decay into the new element. it can also be regarded as the time for the activity to halve, although this can be complicated by the further activity of the new daughter element.

Now compare your solution with the answer below

One possible solution to the paragraph problem above

Radioactivity is the random decay of atoms which change into new elements with the emission of other particles and electromagnetic radiation.

There are three principal emissions: alpha and beta particles, and gamma radiation. Alpha particles are helium nuclei, two protons and two neutrons ejected from the nucleus. The helium nucleus is a particularly stable entity and is ejected as such rather than as the individual nucleons. As a result of alpha emission, the proton number of the original radioactive element drops by 2 and the nucleon number falls by 4.

Beta particles are electrons emitted from the nucleus. The electron appears when a nuclear neutron decays into a proton. The proton number here goes up by 1 and the nucleon number stays constant.

Gamma radiation is emitted as a result of the loss of energy by the new nucleus, normally created in an excited state. Gamma radiation is a high-energy electromagnetic radiation with a short wavelength and high frequency.

Radioactive elements decay with a half-life. This is the time taken for half of the original atoms to decay into the new element. It can also be regarded as the time for the activity to halve, although this can be complicated by the further activity of the new daughter element.

The following scheme is very flexible. Do you only have three main ideas? Then leave out the fifth paragraph. Do you have some subsidiary points to make? Add them like this:

★ introduction
★ new paragraph: main idea 1
    – new paragraph: idea 1a
    – new paragraph: idea 1b
    – and so on …
★ new paragraph: main idea 2
★ new paragraph: main idea 3
★ new paragraph: main idea 4
★ and so on …

The essential point is to produce a logical structure to your work. Having the skeleton in your head to begin with saves time during the examination.

There is always the problem of identifying the main points in the first place. Try a spider diagram as a way of pulling your thoughts together quickly. Once the diagram is drawn, the material is there on paper and there is no danger of you forgetting it again. The examiner will be impressed by your planning so just cross out this rough working neatly, there is no need to erase it completely.

Try to quote equations and relationships where possible, it is also permissible to include estimates of appropriate quantities. Remember to say what your assumptions are.

NOW TRY THIS

Plan this essay

Question (meant to take 45 minutes including preparation time):

Write an essay about waves.

Wave motion occurs in many situations in physics, technology and in nature. Choose **four** examples, as different as possible, that illustrate the great variety of waves which exist. Your answer should include descriptions of these four examples and should show how they differ in origin, physical nature, speed, frequency and wavelength. Mention any features that the various waves have in common.

Four of the 20 marks will be for the quality of your written communication.

20 marks

Hint It is hard to prevent an essay like this descending into a collection of notes – and this is precisely the way to lose the spg marks. You need to get some good physics into the essay, not simply give a catalogue of properties.

The *beginning* of a spider diagram could look like Figure 4.1:

or like Figure 4.2:

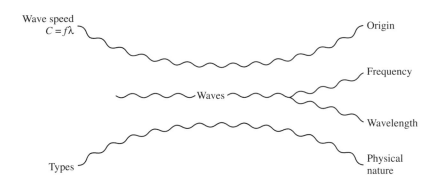

Figure 4.1 Wave spider diagram 1

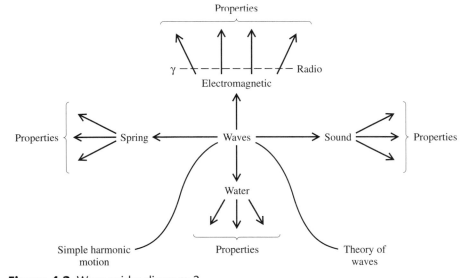

Figure 4.2 Wave spider diagram 2

The second may lead to the better essay as the main ideas will be physics-based rather than focused on examples.

Incidentally, make sure that the waves really are different. Try not to use two electromagnetic waves such as radio and X-rays – there is little to say about the differences.

So a possible essay plan could be:

★ Introduction
  - wave phenomena
  - common properties of waves, $v = f\lambda$ so comments about frequency include wavelength, movement of energy without movement of medium
  - state the four examples you are using

★ Main idea 1: origin and physical nature
   – describe the origin of how the four are made and what their physical nature is

★ Main idea 2: speed
   – typical speed of each of the four

★ Main idea 3: frequency and wavelength
   – mention the ranges of wavelengths that each can have

★ Conclusion
   – perhaps mention the importance of wave motion to us

But this is not the only solution for this question. Try to formulate your own completely different essay structure.

NOW TRY THIS

Plan (but do not necessarily write in full) these essays:

1 Physical laws can be used to make accurate predictions about:
   – the behaviour of gases
   – the decay of radioactive nuclei
   – the thermal transfer of energy.

   In each case it is impossible to predict the behaviour of a single individual component of the system.

   Write an essay about randomness in physics making reference to the basic principles, theory, experimental evidence, and a numerical or algebraic treatment where relevant.

2 There is a considerable body of scientific opinion suggesting that the world must move away from a dependence on fossil fuels.

   Choose **three** alternative sources of energy. For each of these sources discuss:
   – how it can be utilised
   – any environmental consequences that may arise from more use of this source of energy
   – the contribution that it could make to the global energy requirement.

   Support your arguments with estimates where possible.

   Hint In your answer you might consider whether the source of energy you are discussing can be used directly or whether it is more sensible to transform it to a different form and whether it only provides energy for a small community or can be more widely distributed.

3 What is radiation?

   Write an essay explaining carefully the nature of some of the many kinds of radiation that are all around us and the effects that this radiation may have upon people and the environment.

> **Hint** Your essay should contain a discussion of **three** examples of radiation, which differ widely in nature and which are neither all electromagnetic nor all harmful.
>
> **4** Write an essay entitled 'Electrons'.
>
> **Hint** Include in your essay a discussion of conduction by electrons and of the nature of the electron. Describe relevant experimental evidence and introduce equations, relationships and numerical calculations and/or data that help to illustrate your answer.

Practical examination questions

Pieces of writing about practical work (whether as a description of an experiment in a theory paper or arising from a practical examination) can be approached in a very similar way to extended essay work.

The question itself will often help you to structure your answer, if you do not have a structure ready before you go in to the examination room.

> **QUESTIONS**
>
> A simple pendulum consists of a small mass suspended by a light inextensible string from a fixed support. Describe how you would carry out an experiment in order to determine the relationship between the length of a simple pendulum and the periodic time of its oscillation.
>
> **Hint** In your account you should:
>
> ★ provide a clear labelled diagram or a description of the apparatus
>
> ★ indicate the measurements you would make
>
> ★ describe any precautions you take to improve the accuracy of your experiment
>
> ★ show how you would treat the data in order to determine the relationship
>
> ★ state the result that you would expect to obtain.

Answering numerical questions

In many ways, provided that you actually know the physics, these present fewer problems than written answers. There are guidelines to prevent you from making elementary mistakes in Chapter 5.

Some tips for the examination day

If the whole paper is compulsory with no question choice, do the easiest questions first. No-one says you have to answer the paper in the question order. Having said this, the first question is often quite an easy one to help you to settle down. Some specifications try to set an easier final question too. Within this framework, though, questions may be set in specification order.

Research report skills

The requirements of some specifications include a research report as part of the synoptic assessment. Preparing a research report is a task well suited to a test of your synoptic skills because in the course of preparation you will:

★ **use** your knowledge of physics
★ in order to **understand** the topic and then
★ **integrate** the various strands of the topic in order to
★ **write** the report.

So, breaking down the task into its various segments, the task consists of:

★ a choice of a general topic
★ research into the topic
★ understanding the general topic
★ focusing on a specific topic
★ writing a draft report on the topic
★ revising the draft.

Choice

Given that you are writing a report that is intended to demonstrate your synoptic skills, one of your primary needs is for a topic that allows you to draw on various aspects and areas of physics. Start by thinking about the general area of your research as early as possible. Meanwhile decide where the information for the report will come from.

Most people write best when dealing with a topic that stems from their own interests. If astronomy is a particular interest, then feature an astronomical topic. If your hobby is cooking, look at some of the interesting physics involved in food preparation. Your aim is to produce an interesting piece of writing, so make sure that the work – which is likely to take a substantial part of your time for a while – will be of interest to you too.

There is some virtue in choosing an unusual or novel topic. Many physics teachers have read more reports on compact-disc technology than they care to remember, but an interesting and unusual topic immediately grabs the reader's eye and will be favourably viewed.

Make sure that the topic you choose will have plenty of source material. If you try to work from one or two sources, this will be obvious in your work which may tend to develop into a re-hash of someone else's ideas.

You can begin by having a fairly general topic for your initial stages of research. But as you become more and more familiar with the subject matter of your topic and the physics behind it, you should try to make your topic much more specific. Final reports entitled *The physics of bread making* or *The science of modern domestic ovens* will yield much more focused and coherent pieces of writing than *Physics in the kitchen*, even though a broad look at the physics of cookery may have been where you began.

Be clear about the time you will be allowed and the facilities which are available. Will you be given access to libraries (possibly outside school or college) or to the Internet in school time? Or will you have to do all this under your own steam? This will have a bearing on the amount that you can achieve.

Ensure that the physics underpinning the topic is at least of A-level standard. Equally, it must not be so hard that you will struggle to come to terms with it. If you are not especially comfortable with mathematics, then a mathematical topic is inappropriate for you. There are plenty of other topics that you might choose.

Make sure that you know the word limits imposed by the examination rules or by your teachers. Take care to stay within the limits.

Try to obtain sight of the marking criteria (remember that you can access the specifications themselves on the Internet) as this may well change the emphasis of your work. It is of the greatest importance that you should produce a report that allows your teachers to award you the marks available. For example, if you are awarded marks for evaluating the quality of your sources, then your failure to do so will mean that you cannot receive these marks. A scrutiny of the specification should help you to avoid throwing marks away like this. There is more detail about the exercise of scrutinising specifications in Chapter 7, dealing with practical work.

Research

The quality of research that you carry out is very important. This is where you will develop your own ideas about the topic and the main way in which you will come to understand the underlying physics.

Main types of material that you could look at in the course of your research include:

★ books

★ newspaper articles

★ articles from scientific magazines with a general readership e.g. *New Scientist*, *Scientific American* and *Physics World*

★ articles from specialist science papers, published in magazines called journals

★ the Internet

★ radio or television programmes

★ interviews with people you have contacted

★ manufacturer's published data either in printed or Internet form.

The books and magazines can be consulted in libraries. If you live close to a university you may be able to have access to the science library there, but do not expect to be able to take the material away. These libraries can only be used for reference by most people. If you are not near to a major library such as this, your local public library will be able to arrange interlibrary loans for you. But you must apply a long time in advance and know exactly which book you want to borrow; there may be a charge for the service.

Data are often provided readily by manufacturers but, again, you need to write to them very well in advance. Do not expect to telephone the manufacturer and to receive the material by return of post. Remember that the provision of this service is not the primary function of the company.

Interviews can also be very good value. Professional people, doctors, university lecturers, employers are often very willing to talk about their work. They usually see the dissemination of information as an essential part of their job. You will, however, need to approach cautiously usually via a telephone call to a secretary, or a letter sent in good time. Make it clear from the outset what you are doing, indicate the length of time you think that the interview will take, and, if you are asked, supply a list of questions in advance. On the day of the interview, be punctual, dress smartly and stick to the agreed time. Write down the answers or use a tape recorder (with permission). Thank your interviewee at the end. It is a courtesy to follow the interview with a letter thanking the person for his or her time.

Finally, a most important point about your research work and materials that you need to be aware of from the start is the providing of references. Keep full notes, not just of the information *in* the references but also of the references themselves. At the end of your final report you will need to include a bibliography with details of all the sources you use or quote in the report – and this bibliography should be sufficiently detailed for a reader to go back to the original book or article you are using. Fortunately there are standard ways to give these bibliographic references. The following are examples:

★ **A reference to a book**
Name of author(s). Year of publication. *Title of book (in italics)*. Name of publisher, place of publication. Volume number (if more than one), page numbers (if relevant).

So the bibliographic reference to this book is:

Homer, D. 2002. *Synoptic skills in advanced physics*. Hodder and Stoughton, London.

★ **A reference to a scientific article**
Name of author(s). Year of publication. Title of article. *Title of periodical (in italics)*, Volume and part number, beginning and end page numbers.

A reference to a *Physics World* article.

> Bridgman, R. 2001. Guglielmo Marconi: radio star. *Physics World*, Vol. 14, No. 12, pp 29–33.

★ **A reference to a web site**
Details of the web site, web address

> Building atoms from scratch. http://www.pbs.org/wgbh/aso/tryit/atom/

★ **A reference to a CD-ROM**
Title. Year of publication. Publisher, location.

> Advancing Physics AS Students version. 2000. Institute of Physics Publishing, Bristol.

Understanding the topic

As you work at your research you will come to a greater understanding of the general topic you have chosen. You will also become more aware of the specific topic that has most interest for you. This will encourage you to do more focused research on this topic.

You may find that you need to develop your physical understanding even more. This will mean further reference to A-level textbooks and other reference material, perhaps even the Internet.

An important warning

Scientific books and articles are usually very carefully checked before publication – often by anonymous experts in the subject. This is called peer review and is one of the important checks made on published science. No one checks articles on the World Wide Web, however. If you quote from the Internet without checking the information, you do so at your peril.

This can be turned to your advantage. Some boards actively encourage you to criticise and compare sources in your research report. A comparison between reliable sources and ones in which you can have less confidence can be a good solution to this need.

Writing

If at all possible try to do the writing on a word processor. The ease with which text can be written and then re-drafted and revised will make the whole task so much easier. You will also be able to spell check the document and print it out quickly. The processor will take care of routine jobs such as paginating the document and adding neat headers and footers to the text. If you are using a computer BACK EVERYTHING UP every time you make additions or corrections to the text.

The process of writing ought to be:

★ producing an outline

★ writing a first draft

★ revising the draft as often as is needed to convey your ideas in as concise a way as possible.

Revision is a crucial part of the writing process: ask any professional writer. This is the time when you:

★ rearrange the text into its most logical order

★ refine your style of writing

★ check spelling and grammar

★ check for holes in your arguments or errors in the science (perhaps you need more research for some areas)

★ tauten the prose by removing redundant words and phrases.

As you move towards the final version there is much to be said for showing a printed version of your report to a student friend – perhaps you could read your friend's project work too? Ask this reader to be critical and to say which parts (if any) need to be clarified. Ask them also to mark any mistakes in the English on the page margin.

It is important to carry out your own final check on the text on a paper copy. Many people find errors more easily on a printed page than on the computer monitor.

Final points

When the essay is finished and saved on the computer it is time for a final check of the requirements for the format of the printed version – then the essay can be printed out. Use the best quality paper you can and make sure that your printer cartridge is fresh. Diagrams should be on separate sheets of paper from the text and interleaved with it in approximately the correct place. It may not be an effective use of your time to draw the diagrams using computer software unless you are very experienced in using drawing packages. It is better and quicker to draw the diagrams by hand.

Finally, compile and bind the whole report and hand it in on or before the deadline.

CHAPTER FIVE

CONNECTING USING NUMBERS

Physicists model the universe using numbers; mathematics is a crucial tool to enable this. In order to achieve high marks in a physics examination you need to be able to handle calculations and simple mathematical ideas fluently.

This chapter reviews the range of mathematical, graphing and data-analysis skills that are required for synoptic assessment. You will find, however, that the advice here is relevant to your work in other parts of your physics. For example, examiners frequently ask candidates to plot and use graphs in both synoptic and non-synoptic questions.

Chapter 6 deals with the nuts and bolts of drawing and using graphs effectively in your examinations.

Mathematical skills

A list of mathematical techniques that physics students need to know for their A-level examinations is shown below. It is not daunting, you will have met many of these techniques at GCSE.

Mathematical techniques you are expected to be able to understand and use:

Arithmetic and computation

● **Work in decimals and in standard form.**

● **Use ratios, fractions and percentages.**

● **Use calculators to find and use**

$$x^n, \; 1/x, \; \sqrt{x}, \; \log_{10}x, \; e^x, \; \log_{e}x.$$

● **Use calculators to handle sin x, cos x, tan x, x expressed in degrees or radians (degrees are used at AS, degrees *and* radians at A2).**

Handling data

● **Make order of magnitude calculations.**

● **Use an appropriate number of significant figures.**

● **Find arithmetic means.**

Algebra

● **Change the subject of an equation by manipulation of the terms, including positive, negative, integer and fractional indices.**

53

- Solve simple algebraic equations.

- Substitute numerical values into algebraic equations using appropriate units for physical quantities.

- Understand and use the symbols $=$, $<$, $<<$, $>>$, $\propto$, $\sim$.

Geometry and trigonometry

- Calculate the:
  - area of triangles
  - circumferences and areas of circles
  - surface areas and volumes of rectangular blocks
  - area of cylinders
  - area of spheres.

- Use Pythagoras' theorem and the angle sum of a triangle.

- Use sines, cosines and tangents in physical problems.

- (*A2 only*) Understand the relationship between degrees and radians and translate from one to the other.

Graphs

- Translate information between graphical, numerical and algebraic forms.

- Plot two variables from experimental or other data.

- Understand that the equation $y = mx + c$ represents a linear relationship.

- Determine the slope and intercept of a linear graph.

- Draw and use the slope of a tangent to a curve as a measure of rate of change.

- Understand the possible physical significance of the area between a curve and the *x*-axis and be able to calculate it or measure it by counting square as appropriate.

- (*A2 only*) Use logarithmic plots to test exponential and power law variation.

- (*A2 only*) Sketch simple functions including $y = \frac{k}{x}$, $y = kx^2$, $y = k/x^2$, $y = \sin x$, $y = \cos x$, $y = e^{-kx}$.

Tips for answering numerical questions

Write it all down

Some of the equations that you need to answer a question may be on the list of relationships you are **required** to know (see opposite). In this case there may well be a mark available simply for quoting this equation – do not forget to do so even though you may think that both the use of the equation and the equation itself are obvious. You may be provided with a list of other equations as part of the examination paper – often on a tear-off sheet. Again, do make it clear to the examiner what you are doing in your calculation, what numbers you are substituting and how you arrived at your answer.

Relationships you are required to know by heart

- Speed = distance/time

- Force = mass × acceleration

- Acceleration = change in velocity/time

- Momentum = mass × velocity

- Work done = force × distance moved in direction of force

- Power = energy transferred/time taken = work done/time taken

- Weight = mass × gravitational field strength

- Kinetic energy = ½ × mass × speed²

- Change in potential energy = mass × gravitational field strength × change in height

- Pressure = $\dfrac{\text{force}}{\text{area}}$

- Pressure × volume = number of moles × molar gas constant × absolute temperature

- Charge = current × time

- Potential difference = current × resistance

- Electrical power = potential difference × current

- Potential difference = energy transferred/charge

- Resistance = resistivity × length/cross-sectional area

- Energy = potential difference × current × time

- Wave speed = frequency × wavelength

- Centripetal force = mass × speed²/radius

- $F = \dfrac{Q_1 Q_2}{4\pi \varepsilon_0 r^2}$

- $F = \dfrac{G m_1 m_2}{r^2}$

- Capacitance = charge stored/potential difference

- Potential difference across coil 1/potential difference across coil 2 = number of turns in coil 1/number of turns in coil 2.

If you find learning these difficult try to use mnemonics (memory aids). One trick is to use the triangle idea, Figure 5.1, so that **Force = mass × acceleration** becomes:

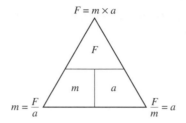

$$F = m \times a$$

$$m = \frac{F}{a}$$

$$\frac{F}{m} = a$$

Figure 5.1 Force, mass, acceleration triangle

Or try making the equation into a word or phrase.

If you make them up yourself, you are more likely to remember them.

Express the answer properly

An easy way to lose marks is to be sloppy about the way you express the final answers to numerical questions. Algebraic answers do not normally require a unit. Get into the habit of *automatically* checking three things about every answer.

Checking numerical answers

Check 1. Have you given a unit?

Only a few quantities in physics never need a unit added to them. They include any ratios and any quantity derived from a ratio, e.g. energy efficiencies.

Always try to learn the appropriate units for quantities when you meet them, and make a special point of including them in your revision notes.

Check 2. Is the unit correct?

It is easy to make a careless slip with a unit. Take extra care and do not lose your understanding and knowledge of the subject under pressure. A calculation of the rate of change of energy with time has a unit of joules per second – so write $J\ s^{-1}$ or (better) W, not simply J.

Check 3. Have you expressed the number to the correct number of significant figures?

This rule is simple: **always express the answer to the same number of significant figures as those in the question.**

For example

Question: Calculate the speed reached by a motor car accelerating from rest at $2.5\ m\ s^{-2}$ for 9.0 s.

Answer: $v = u + at$; $v = 0 + 2.5 \times 9.0 = 22.5$ m s^{-1}

Round this answer to 23 m s^{-1} to agree with the figures in the question.

It is reasonable to show the intermediate answer to convince the examiner that you have done the working correctly but make the final answer conform to this rule.

BUT THERE IS AN IMPORTANT EXCEPTION

One type of question breaks this rule. The mark scheme for a 'Show that ...' question will usually require you to go to one **more** significant figure than the suggested answer to prove that you have actually carried out the calculation.

For example

Question: Show that a motor car accelerating from rest at 2.5 m s^{-2} for 9.0 s reaches a speed of about 20 m s^{-1}.

Answer: $v = u + at$; $v = 0 + 2.5 \times 9.0 = 22.5$ m s^{-1}

Quote this answer as 22.5 or 23 but **not** 20.

Powers of ten

Another way to lose a mark through a careless mistake is to quote powers of ten incorrectly.

Answers appear on most calculators in the form:

$$1.05 \; {-}02$$

whereas what the calculator is really showing is:

$$1.05 \times 10^{-2}$$

Take great care to express the numbers correctly. Examiners are often instructed to penalise those candidates who clearly do not understand the numbers that they are writing down.

Using the calculator reliably

If you are lucky enough to own a good calculator, make sure that you know how to use it properly. There are some very sophisticated machines on the market that will – for the price of a few pounds – carry out many different types of calculations for you. But are you one of those people who only ever use the $+ - \times$ or $\div$ keys?

Perhaps your calculator can work out averages quickly? Some calculators can evaluate the gradient and position of a best-straight line, can yours? Do you know how to do so? Calculator instruction manuals are the most under-used resources in physics laboratories today.

NOW TRY THIS

Go back to the list of mathematical requirements on pages 53–54.

Look at the list of things you need to be able to do with your calculator, do you know where *every* key is for these?

Now look at the rest of the techniques again:

Make a list of all the ways your calculator might help you (degrees to radians, means, plotting best-straight lines on graphs). Ensure that you really can do these quickly and reliably.

Prefixes

By all means use prefixes and powers of ten to avoid the clumsy appearance of large numbers of zeros, but do use them correctly. Look at Table 5.1 to see the most commonly used ones at A-level (there are others allowed in the prefix system).

Prefix	Symbol	Factor
giga	G	$\times 10^9$
mega	M	$\times 10^6$
kilo	k	$\times 10^3$
deci	d	$\times 10^2$
centi	c	$\times 10^1$
milli	m	$\times 10^{-3}$
micro	μ	$\times 10^{-6}$
nano	n	$\times 10^{-9}$
pico	p	$\times 10^{-12}$

Table 5.1 Prefixes for powers of ten

For example

The speed of light is 300 000 000 m s^{-1} – do not use large numbers of zeros like this, one of the following options is better:

★ *the speed of light is 3×10^8 m s^{-1}*
★ *the speed of light is 300 Mm s^{-1}*
★ *the speed of light is 0.3 Gm s^{-1}*

Estimates

What is an estimate?

Some quantities in physics can only be estimated, usually because some of the variables involved in the calculation are not known exactly. In this case the physicist has to make an educated *guess* at the value of one or more variables in the calculation.

Estimating from scratch

Here is an example of an estimate for you to try; there are hints to get you started. When answering an estimation question, remember to make it clear what quantities you have estimated and what values you actually knew before you began.

NOW TRY THIS

A man jumps off a wall. Estimate the force that he exerts on the ground as he lands.

Hint Guess the height of the wall from which it would be safe to jump then calculate the landing speed, you know the acceleration due to gravity ($9.8 \ \text{m s}^{-2}$); estimate how long the process of landing takes; estimate the man's mass and calculate his change in momentum; use

$$force = \frac{change \ in \ momentum}{time \ taken \ for \ change}$$

to work out the force.

Remember to use kilograms, seconds, metres as usual. Quote the answer to a sensible number of significant figures. After all, you have guessed almost everything in the question!

NOW TRY THIS

1 What is the total weight of the house in which you live?

2 How many key depressions were involved in typing out the manuscript for this book?

3 What is the total floor area of your school?

4 What is the total length of wire in a piano?

Uncertainties

This section on experimental uncertainty is included here as a mathematical aspect of A-level physics, but it could have easily found its way into the chapter on practical examinations and investigations. You may wish to read it in conjunction with Chapter 7 and also when you are carrying out practical investigation work.

Any measurements are subject to error. There are two types: **systematic** errors and **random** errors.

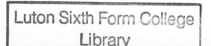

Systematic errors

Imagine an old wooden metre rule with 2 mm of material worn away from the zero end. Anything that is measured with this rule will be measured as too large – 2 mm too large to be exact. This is a systematic error. Another example is an analogue electrical ammeter where the needle does not point to zero when no charge flows through it. All measurements will be in error by a constant amount. Repetitions of the measurement will not eliminate the error.

Random errors

Imagine a metre rule (a full 1000 mm this time) which is used to find the length of a piece of paper. The measurement is carried out carelessly with the rule held well above the paper. The chances are that there will be a substantial parallax error (see Figure 5.2) between the observer's eye, the rule and the paper.

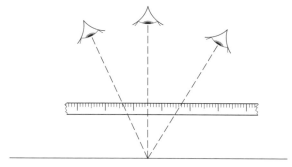

Figure 5.2 How parallax error arises

Many repetitions of this measurement and a subsequent averaging will produce a value close to the correct paper length.

The repetition of measurements and the averaging of the results can therefore help with random errors. Systematic errors, on the other hand, must be eliminated by other means.

Uncertainty in measurements

A good rule of thumb is that the uncertainty in a single measurement is equal to the size of the smallest scale division. So the measurement made by a metre rule of the long side of a piece of A4 paper is 293 ± 1 mm. The short side is 210 ± 1 mm.

If you are taking averages of a number of readings to give a mean value for a measurement, the appropriate value to use is the standard error. Many modern calculators will work this out quickly. But another quick and perfectly adequate estimate of the standard error is to find the difference between the largest and smallest value and then to divide this difference by the number of measurements.

> ### EXAMPLE
>
> The length of the long side of an A4 page was measured in millimetres, with a half-metre rule, to be:
>
> 297 298 295 296 295 294 296 296 298 298 297 297 299
>
> The average is 296.6. The highest value is 299, the smallest is 294 so the standard error is 5/13 = 0.38.
>
> Express the answer as 296.6 ± 0.4 mm. This statement means that the actual answer is likely to lie somewhere between 296.2 mm and 297.0 mm.

Combining errors

As well as estimating the error in a measurement, you can be expected in some examinations to be able to combine errors. There are three cases:

★ adding or subtracting measurements
★ multiplying or dividing measurements
★ raising measurements to a power.

Combining errors when adding or subtracting

Suppose you have measured all sides of an A4 sheet of paper and you want to calculate the circumference. Figure 5.3 shows the four measurements:

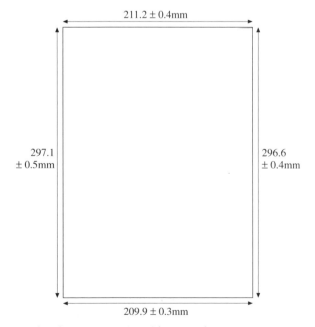

Figure 5.3 An example of measurement with quoted error

The smallest value that the circumference could be is obtained by taking the four smallest estimates of the lengths, that is 296.2 + 296.6 + 210.8 + 209.6 = 1013.2 mm.

The largest value is 297.0 + 297.6 + 211.6 + 210.2 = 1016.4 mm.

So the average of these two values is 1014.8 ± 1.6 mm.

But notice that the error of 1.6 is actually the sum of the errors 0.4 + 0.5 + 0.4 + 0.3 = 1.6 mm.

So the rule is: **when adding or subtracting errors add the errors** – these are usually known as absolute errors.

Combining errors when multiplying or dividing

Suppose that you now want to find the area of the A4 sheet using the measured quantities 296.6 ± 0.4 × 211.2 ± 0.4.

You cannot use absolute errors this time. First, calculate what is known as the fractional uncertainty for both errors.

$$\text{Fractional uncertainty} = \frac{\text{absolute error}}{\text{value of the measurement}}$$

The fractional uncertainty in the measurement of the long side is 4 parts in 2966 or 1 part in 741.5, 1/741.5 = 0.001348.

The fractional uncertainty in the measurement of the short side is 4 parts in 2112 or 1 part in 528, 1/528 = 0.001894.

The area of the A4 sheet is 296.6 × 211.2 = 62641.92 mm^2.

The fractional error of this answer is the sum of the two fractional errors concerned.

Fractional errors of the answer = 0.001348 + 0.001894 = 0.003242.

Now turn the fractional error back into an absolute error.

0.003242 of 62641.92 is 203.08510464. This calculation has been carried through with all its significant figures so that you can check what is going on. Of course, not all of these are needed.

The answer could be expressed as 62641 ± 203 mm^2 but there is almost no meaning to the last two digits either in the answer or the error figure and so the best way to write the answer is as:

$$(626 \pm 2) \times 10^2 \text{ mm}^2$$

So, the rule now is: **when multiplying and dividing add the fractional errors** (not forgetting to convert them back to absolute errors at the end).

Combining errors when using powers

This is an extension of the previous rule for multiplication and division.

Suppose that the sheet of paper is now square with a side of 296.6 mm. The area would be $296.6^2 = 296.6 \times 296.6 = 87971.56$ mm^2.

The fractional error of this answer is obviously the sum of the two fractional errors (which are the same) because the square is actually the two numbers multiplied together.

So the rule is: **in raising an error to a power, multiply the fractional error by the power.**

In this case the fractional error of the answer is $2 \times 0.001348 = 0.002696$.

And the absolute error is $0.002696 \times 87971.56 = 237.17132576$, the final answer being:

$$(879 \pm 2) \times 10^2 \text{ mm}^2$$

NOW TRY THIS

1   Figure 5.4 shows the readings on a voltmeter connected across a resistor and on an ammeter in series with the resistor. Read the meters and add estimates of the errors in these values. Calculate the resistance of the resistor including the error in the value.

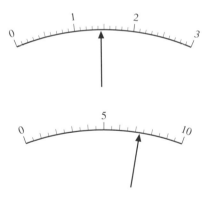

Figure 5.4 Voltmeter and ammeter readings

2   A simple pendulum is used to make a calculation of g, the acceleration due to gravity. The equation for this determination is $T = 2\pi\sqrt{(l/g)}$, where T is the time period of the pendulum and l is the length of the pendulum.

In the experiment, 25 oscillations took 40.4 ± 0.1 s, when the length of the pendulum was 0.650 ± 0.001 m.

Calculate the value of g including an estimate of the error.

Analysing using graphs and data

Graphs

A curve shows the basic change of one variable with respect to another.

A straight line helps you to see the exact nature of the relationship between the variables.

Once you have a straight-line graph you can make precise mathematical statements about how the quantity on the y-axis varies with respect to the quantity on the x-axis, in other words to write an equation describing the relationship between y and x. The moral is: always manipulate your data to produce a straight line if you can.

If the question tells you what to plot, make the most of this information and make sure that you do what was asked.

Sometimes, though, the problem is deciding what to plot. Figure 5.5 reminds you of the equation of a straight line.

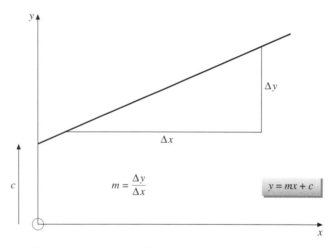

Figure 5.5 A straight-line graph related to its equation

EXAMPLE

Deciding what to plot

Step 1: take the physics equation you know.

Step 2: manipulate the algebraic equation so that it is in the form $y + mx + c$.

Step 3: plot the graph.

EXAMPLE *continued*

Example 1

A stretched guitar string with a length *l* will emit a sound of frequency *f* when the tension in the string is *T* and the mass per unit length (the mass of a metre length) is *m'*. The equation for this is:

$$f = \frac{1}{2l} \sqrt{\frac{T}{m'}}$$

(i) In an experiment a student measures the frequency whilst varying the length of the string. What graph should be plotted to give a straight line? *T* and *m'* (and ½) are constants in this experiment, so $f = \left[\frac{1}{2} \sqrt{\frac{T}{m'}} \right] \frac{1}{l}$.

The graph to plot is *f* (*y*-axis) against $1/l$ (*x*-axis), this will be a straight line with a gradient of $0.5 \sqrt{\frac{T}{m'}}$. If there are no systematic errors, it ought to go through the origin (Figure 5.6(i)).

(ii) The student goes on to measure the frequency for differing values of the tension in the string. The length and the mass per unit length are constant this time. So, rearranging gives:

$$f^2 = \left(\frac{1}{4m'l^2} \right) T.$$

A plot of f^2 against *T* will give the straight line with a gradient of $\frac{1}{4m'l^2}$ (Figure 5.6(ii)).

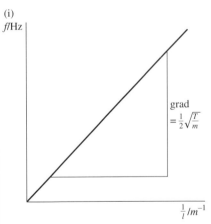

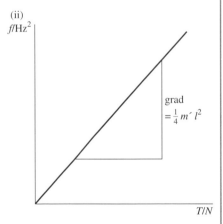

Figure 5.6

Example 2

Question: A cell with an emf of ε volts and an internal resistance of *r* ohms is supplying current to a resistor. In the experiment the potential difference across the resistor *V* and the current *I* are being measured. What graph will give a straight line?

Answer: The equation here is:

$$V = \varepsilon - Ir.$$

EXAMPLE *continued*

Re-arrange to:

$$V = -rI + \varepsilon$$

and compare this with:

$$y = mx + c.$$

A plot of V (y-axis) against I (x-axis) will give a straight line of gradient $-r$ (the internal resistance) with an intercept on the y-axis equal to the emf of the cell (Figure 5.7).

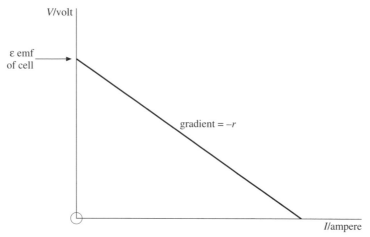

Figure 5.7

NOW TRY THIS

1 A student investigating the change in length of a piece of wire applies a force F to the wire and measures the new length l of the wire. The starting length of the wire is l_0. Sketch the graph that the student will expect to plot.

Hint force is directly proportional to extension; extension is the difference between the original length and the present length.

2 A student measures the pressure p of a gas at a series of volumes V. What graph will yield a straight line. Sketch the graph and state the gradient of the line.

Hint $pV = nkT$.

3 A trolley is moving with a constant acceleration a. A student begins to measure the distance travelled s at a time t when the speed is u. Sketch the graph that will give a straight line.

Hint remember that $s = ut + \frac{1}{2}at^2$; divide every term by t to eliminate the t and t^2 problem.

Using logarithms in graphs

(i) When you know the relationship is exponential to begin with

A-level physics contains a number of examples in which two variables are connected by an exponential relationship. They include the discharge of capacitors, the decay of radioactive atoms and the way in which damped oscillators change the amplitude as time goes on. Thermally activated processes (such as viscosity and the rate at which many reactions occur) also vary with temperature in a way that is exponential.

In radioactive decay, the number of atoms left, N, depends on the decay constant λ, the time t and the original numbers of atoms N_0 as:

$$N = N_0 e^{-\lambda t}.$$

The graph of N against t looks like Figure 5.8a.

To produce a straight line, first of all take logs of both sides of the equation:

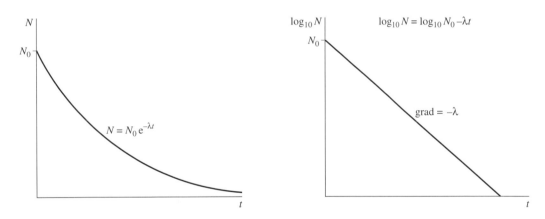

Figure 5.8 An exponential relationship plotted in two ways

$$\log_{10} N = \log_{10} N_0 - \lambda t \qquad \text{or} \qquad \log_e N = \log_e N_0 - \lambda t$$

(Notice that for this exercise, producing a straight line, it does not matter whether you use logarithms to base 10 or to base e.)

Now compare this with the equation of the straight line $y = c + mx$.

Figure 5.8b shows the graph of $\log N$ against t.

N O W   T R Y   T H I S

What will the gradient and intercepts be for these other examples of exponentials?

1   The charge Q on a capacitor as the capacitor (value C, initial charge Q_0) discharges through a resistance R.

2   The amplitude A of a damped oscillator that loses a fraction F of its energy each oscillating cycle, where n is the number of cycles.

(ii) When you are not sure of a power law

This is a technique to use in a practical examination or in practical project work when you know that the equation is something like:

$$y = kx^n$$

The context is that you have collected a set of data but that you do not know the value of n or k.

This is where a log–log graph comes into its own. By taking logs of both sides of the equation, it becomes:

$$\log_{10} y = \log_{10} k + n \log_{10} x.$$

As usual, rearrange this as:

$$\log_{10} y = n \log_{10} x + \log_{10} k$$

and compare it with:

$$y = mx + c.$$

You should now spot that a graph of $\log_{10} y$ against $\log_{10} x$ has a gradient of n – the power you need to find – and the intercept on the y-axis is $\log_{10} k$.

EXAMPLE

Figure 5.9 shows some data that obey a power law $y = kx^n$,

The log–log graph is a straight line with a gradient of 1.5. The intercept on the y-axis is about 0.4 which is $\log_{10} 2.7$.

So the equation of the line is $y = 2.7x^{1.5}$.

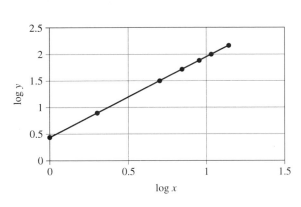

Figure 5.9 Log *y* plotted against log *x* for the data below

y	2.7	7.6	30	50	73	99	140
x	1	2	5	7	9	11	14

NOW TRY THIS

For each of these sets of data find out the value of *k* and *x* in $y = kx^n$.

(i)

y	25	54	77	98	116	134	151
x	1	3	5	7	9	11	13

(ii)

y	2.24×10^3	2.71×10^3	4.08×10^3	7.67×10^3	1.29×10^4	2.29×10^4
x	20	22	27	37	48	64

(iii)

y	0.052	0.089	0.104	0.144	0.169	0.205
x	0.2	0.45	0.57	0.94	1.2	1.6

Data analysis using tables

What data tables can show

Tables help us in two ways: they are a convenient method of storing data in reports to which other people may need access and they are sometimes a useful way of making a point.

Some daily newspapers contain a complete listing of the current value of shares in the London financial markets. No-one could argue that these are easy reading, the data are in small print and contain abbreviations and numbers that are difficult for anyone but financial experts to understand. They are examples of the first type of table: the data store. You may well use a table in this way to display your results in an appendix if you have to write an investigation report as part of your coursework for A2.

The same newspapers will use tables in the second way too – to convey one or more points quickly to the reader. Tables for this purpose will generally be short, visually appealing and contain an obvious message.

If you are using tables, be clear what you need your table to do for the reader.

Displaying data well

Every row or column should be clearly labelled with the quantity and the unit: electric current/A, speed/m s^{-1}, etc.

Consider the format of your numbers. If possible use scientific (standard) form and place the powers of ten at the head of the row or column.

Thus:

Aircraft speed/m s^{-1} × 10^2	Meaning
1.2	120 m s^{-1}
2.6	260 m s^{-1}
3.7	370 m s^{-1}

Table 5.2

Making things obvious

If you are preparing a research report or a write-up of a practical investigation, you will almost certainly need to display numerical data in some way. The use of a graph or a data table will probably be your first thought. These are of course not the only options. Have you considered other forms of data display?

If you want some further help with the enormous range available take a look at the options available in a computer spreadsheet package such as *Microsoft Excel©*. Programs such as this will take numerical data and display it for you in a variety of different ways (there are equivalent packages by other companies and for other computers). Your task is to select the most appropriate display and use this in your report.

You may be learning communication and IT skills at school or college. Use the information you learn there to improve the impact and appearance of documents in all your A-level subjects.

Testing for relationships

A major use of the data table is to look for relationships between the variables displayed. As a reminder, you are expected to be able to sketch graphs of the following:

$y = k/x$, $y = kx^2$, $y = k/x^2$, $y = e^{-kx}$ together with $y = \sin x$, $y = \cos x$

and it is reasonable for you to be asked to decide whether data in tabular form obey the first four relationships as well.

The trick is to combine the two variables in the relationship in a way that should lead to a constant value if the relationship is correct.

Here is an example:

EXAMPLE

Table 5.3 shows values of pressure and volume for a gas experiment in which the temperature and the mass of gas were kept constant. Is Boyle's law confirmed for this data?

Pressure/kPa	115	110	105	100	95	90	85	80
Volume/m³	15.6	16.2	17.2	18.0	18.8	19.8	21.0	22.3

Table 5.3

Boyle's law states that the pressure p in the gas is inversely proportional to the volume v.

Algebraically:

$$p \propto 1/v$$

if the relationship is true then pv will be a constant.

Add another row to the table (Table 5.4) and add the product of pressure × volume for each pair of data to this row.

Pressure/kPa	115	110	105	100	95	90	85	80
Volume/m³	15.6	16.2	17.2	18.0	18.8	19.8	21.0	22.3
Pressure × Volume/kJ	1794	1782	1806	1800	1786	1782	1785	1784

Table 5.4

Although pV is not exactly the same value for each pressure point, it only varies by 1.3%. So this is good enough to confirm the relationship.

This technique works well for $y = k/x$, $y = kx^2$, $y = k/x^2$.

Table 5.5 summarises the tests

$y = k/x$	test by checking that	$y \times x$ is constant
$y = kx^2$	test by checking that	$y \div x^2$ is constant
$y = k/x^2$	test by checking that	$y \times x^2$ is constant

Table 5.5

$y = e^{-kx}$ needs a different approach. You will probably remember that this exponential relationship leads to half-life behaviour in radioactive decay and capacitor discharge.

The problem with the data-table approach as opposed to the graphical treatment is that you are unlikely to be given (or to collect) consecutive data values that differ by a factor of 2. In any event, you need to perform the check as many times as possible, not just once.

For any exponential change, a change by the same value on the x-axis always leads to identical ratios of y-values.

To illustrate this point, Table 5.6 shows data for the discharge of a capacitor with time.

Charge on capacitor/µC	6.1	3.7	2.2	1.4	0.8	0.5
Times/s	10	20	30	40	50	60

Table 5.6

To test whether this is exponential without resorting to a graph, extend the table by one row (Table 5.7). Notice that the times go up in equal steps – this is important for this test to work. Now calculate the ratio of each charge divided by the charge that was stored 10 s earlier. So the first entry will be at 20 s (there is no 10 s entry because there is no 0 s entry to use for its calculation). The value is $3.7/6.1 = 0.607$. Now repeat this for each time from 30 s onwards.

Charge on capacitor/µC	6.1	3.7	2.2	1.4	0.8	0.5
Times/s	10	20	30	40	50	60
		0.607	0.607	0.607	0.607	0.607

Table 5.7

The ratios are identical: the relationship is exponential.

Just to convince you of this, the graph is plotted as well (Figure 5.10). Does this graph have a half-life property?

Hint compare the time to drop from 6 µC to 3 µC with the time to go from 4 µC to 2 µC.

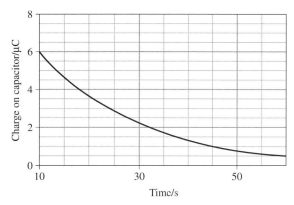

Figure 5.10 Discharge of a capacitor with time

CONNECTING USING DATA AND GRAPHS

Graphs help you to visualise numerical data. You can be asked to **draw** a graph from data that has been given to you in a theory paper or from data you have measured yourself during a practical examination. You may also be asked to **sketch** a graph to illustrate the variation of one variable with another.

Drawing a graph involves taking a set of data and plotting it on axes that have scales, labels and units. You will normally draw a best-fit line or curve through the data (see Figure 6.1).

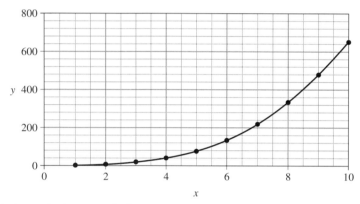

Figure 6.1 A graph drawn with data points

Sketching a graph means producing a drawing that usually has no data points, axis scales or units, but which shows how the variables y and x are related (see Figure 6.2).

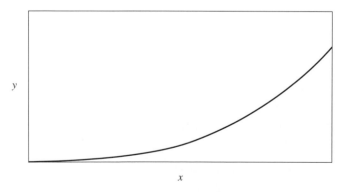

Figure 6.2 A sketch graph that shows how y relates to x

Sketch graphs should

● Be neatly drawn in pencil.

● Show the relationship between the *y* variable and the *x* variable clearly – exaggerate if necessary, but try to get the correct shape.

● Have both axes labelled with the correct variable (with no data points on the graph no units or scale is needed).

● Include any known quantities on the graphs. In this case a unit and markings on the scales *are* required.

Shapes are important

Remember that there are six relationships that you have to be able to **sketch**:

$y = k/x$, $y = kx^2$, $y = k/x^2$, $y = \sin x$, $y = \cos x$, $y = e^{-kx}$.

Can you draw these shapes freehand?

Sines/cosines

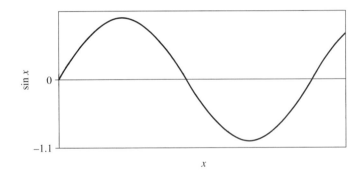

Figure 6.3 Sine curve

Examiners expect to see a reasonable sketch, sine curves as semi-circles are not acceptable.

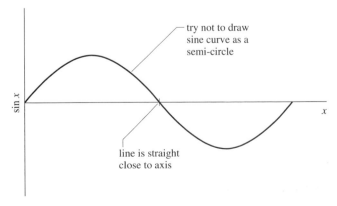

Figure 6.4

$y = 1/x$

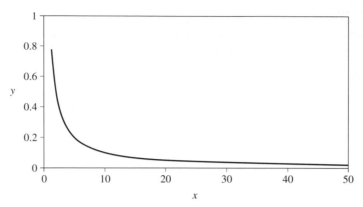

Figure 6.5 $y = 1/x$

$y = 1/x^2$ is steeper than $y = 1/x$

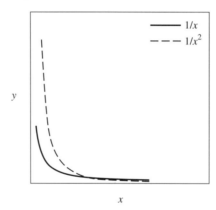

Figure 6.6 Comparing $y = 1/x$ with $y = 1/x^2$

$y = e^{-kx}$

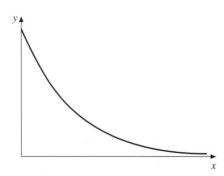

Figure 6.7 $y = e^{-kx}$ crosses the y axis

these cross the y-axis but are asymptotic on the x-axis.

You are, however, very unlikely to be asked to sketch these graphs in this way. The graphs will be set in the context of a real piece of physics.

Question 1: A horizontal force of 1.5 kN acts on a motor car of mass 850 kg that is initially at rest. Sketch graphs of acceleration, speed and distance travelled for the first 15 seconds of the car's motion. You should include labels for the axes and any numerical values that you can evaluate.

Hint   you need first to calculate the acceleration of the car (use $F=ma$); then work out the speed at 15 s – remember that the initial speed is zero; finally you can use $s = ut + \frac{1}{2}at^2$ to find how far the car has gone during the 15 seconds. Your knowledge of physics should tell you the *shape* of the lines on the graphs. The acceleration is constant, the speed therefore varies linearly and the distance is proportional to time squared.

Question 2: The drag force F on a racing car depends on the speed v as $F = v^2$, sketch a graph showing how F varies with v.

[Another set of variables that obeys this relationship is the electric power P plotted against electric current I delivered to a constant resistance.]

Without referring to the graphs above sketch graphs of each of the specified relationships:

$y = k/x$, $y = kx^2$, $y = k/x^2$, $y = \sin x$, $y = \cos x$, $y = e^{-kx}$

For each of the graphs suggest **two** sets of variables that obey the relationship you have drawn.

Drawing graphs

Here are some basic instructions for producing neat, clear graphs from data that will satisfy the requirements of examiners.

Drawing graphs in examinations and coursework

- Use sensible scales 1:2, 1:5, 1:10 **not** 1:3, 1:6, 1:7 or 1:9.

- Fill the graph paper, most examiners expect you to fill more than half the grid provided; this means that you should …

- … use a false origin in order to do this (one that does not start at 0,0).

- Label axes properly with quantity + unit + power of ten e.g. speed/10 m s^{-1}.

- Mark data points consistently and prominently, use $\times \odot +$, use colour to differentiate between different sets of data on the same axes.

- Draw the lines and points with a sharp pencil, examiners will expect your lines to be thinner than the thickest grid lines on the graph paper.

- Use a ruler to draw all the straight lines (see below for how to select a best-fit line).

- Use free-hand for curves. Draw the curve in one flowing movement, having practised the curve (pencil above the paper!) several times first, keep your drawing hand **inside** the curve.

Drawing a best-fit line on your graph
Quickly

- Use a transparent ruler if possible.

- Get a balance of points (the same number) each side of the line. Minimise the total distance from all points to the line.

A common error is to force the line through the origin. Never distort your data in this way unless there is a very good reason.

With more time in hand

Many modern calculators can determine the exact position of a best-fit line. If yours can, learn how to do this. Most A2 graphs have seven data points or fewer and the process is a quick one.

An approximate way to do this is to:

(i) Divide your data into a lower and an upper half (if there is an odd number of data points, make one of the groups one larger than the other; it does not matter which).

(ii) Find the mean (average) of all the x values in the lower set and the mean of all the y values in the lower set. Plot this new point on the graph (use a different plotting symbol so that it is obviously **not** a real data point).

(iii) Do the same thing with the xs and ys from the upper set of data.

(iv) The best-fit line will be close to the line joining the two mean points (Figure 6.8).

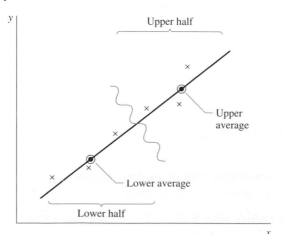

Figure 6.8

Getting data from graphs

Sometimes you need to take data from a graph – either one that you have
plotted or a graph given to you. The techniques available to you are:

Interpolation

Here you simply read off the values on the axes for specified points that you
require.

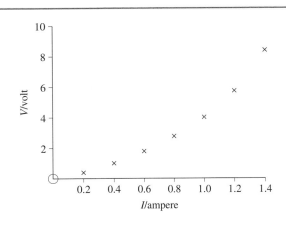

Figure 6.9

Figure 6.9 shows a graph of potential difference against current for a metal wire.
Draw an appropriate trend through the data points and use the graph to calculate
the value of resistance at a current of 1.15 A.

Extrapolation

Whereas interpolation is relatively straightforward, extrapolation is risky. You
have to assume that the behaviour of the graph continues in the same way and
this may or not be the case. Extrapolation occurs only rarely in theory papers,
but you may need to extrapolate values in your own practical work.

The data for another potential difference against current graph are given in Table 6.1:

I/A	0.2	0.6	1	1.4	1.8
V/V	0.2	0.9	2	3.5	5.4

Table 6.1

Estimate what the potential difference across the wire will be when the current is
(i) 0.05A; (ii) 6 A.

State and explain which of these two estimates is likely to be the more reliable?

Gradients

You need to be able to calculate gradients accurately and quickly. This ability will almost certainly be tested at some point in your examinations.

Calculating the gradient of a straight line

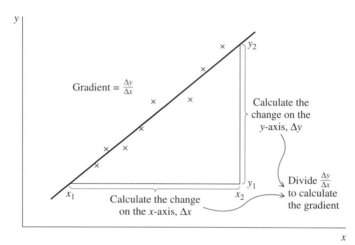

Figure 6.10 How to calculate the gradient of a line

● Draw a large triangle (in an examination, triangles with an hypotenuse **smaller than half the trend line** are penalised). Make your triangle obvious to the examiner (even if you read the values off the scale and do not use the triangle as such) (Figure 6.10).

● Only use actual data points if they lie **exactly** on the line.

● Calculate the change on the *y*-axis, $\Delta y = y_2 - y_1$

● Calculate the change on the *x*-axis, $\Delta x = x_2 - x_1$

● Divide $\dfrac{\Delta y}{\Delta x}$.

● The units are (units of *y*) (units of *x*)$^{-1}$ and they must **not** be omitted.

Calculating the gradient at a point on a curve

This gives the instantaneous value at the point concerned.

Draw a tangent to the curve at the point concerned (Figure 6.11).

If you find this difficult in a practical test, use a mirror as in Figure 6.12.

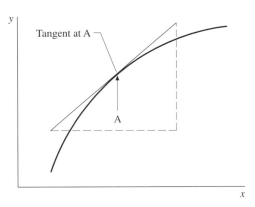

Figure 6.11 A tangent to a curve

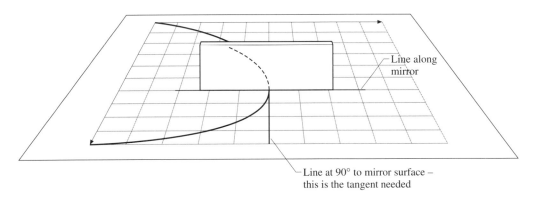

Figure 6.12 Using a plane mirror to draw a tangent

When the curve appears continuous, the mirror is at 90° to the tangent. Draw along the mirror surface and then use a protractor to draw another line at 90° to this.

Now continue as though for a gradient of a straight line.

Intercepts

These usually involve just a read-off on the graph, but if you have used a false origin to make the graph fit the graph paper this may take a little longer.

Figure 6.13 shows a quick way to work out an intercept.

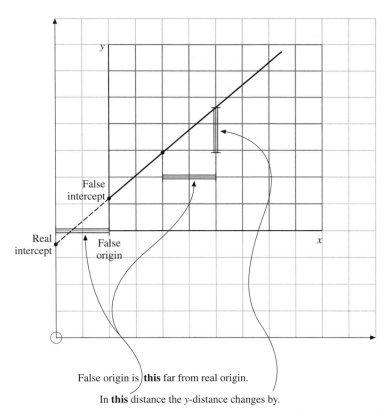

False origin is **this** far from real origin.

In **this** distance the *y*-distance changes by.

So the real intercept must be the **same** *y*-distance below the false intercept.

Figure 6.13 A quick method to calculate and intercept when the origin is false

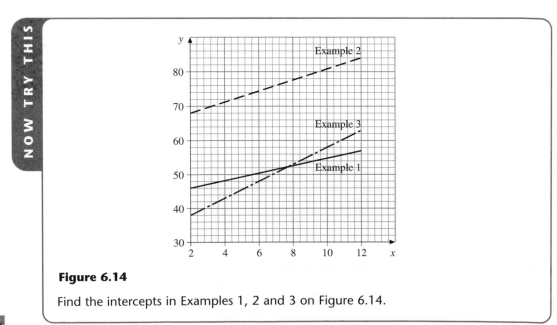

Figure 6.14

Find the intercepts in Examples 1, 2 and 3 on Figure 6.14.

Calculating areas under a graph line
Areas under curved trends

If you are expected to work out the area under a curved trend then the graph will be drawn on a grid. The procedure is straightforward (Figure 6.15):

★ Calculate the area of one grid rectangle.

★ Count up the total number of grid rectangles for the part of the graph you are working on – you may need to estimate the size of grids that are not complete.

★ Multiply the size of one grid rectangle by the total number of grid rectangles.

★ Remember to include the units (*y*-axis unit multiplied by *x*-axis unit).

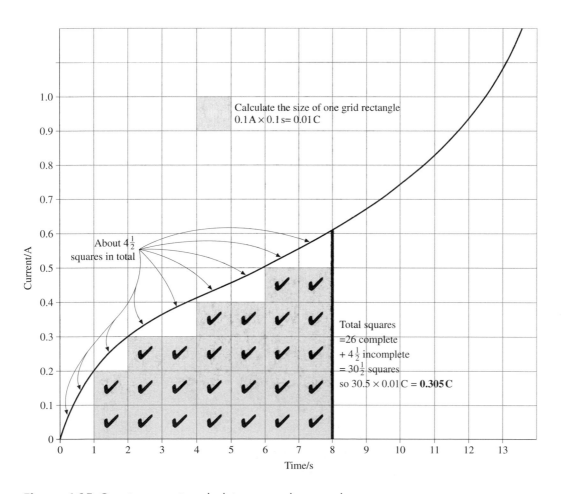

Figure 6.15 Count squares to calculate area under a graph

Areas under straight trends

These are even easier! If you think of the area as being made up of a rectangle and a right-angled triangle, then you should have no problems (Figure 6.16).

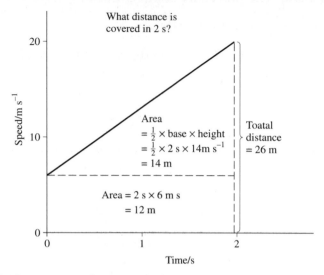

Figure 6.16 Divide an area under a straight line into regular shapes

CHAPTER SEVEN

MAKING PRACTICAL CONNECTIONS

Practical investigations

A practical investigation is a piece of research or project work carried out with apparatus in the laboratory over a period of perhaps two weeks of school time. During this time you will devise experiments, assemble or construct the apparatus, and collect data and carry out an analysis. Finally you will write a project report outlining your work, your analysis and your conclusions.

Three boards set practical investigations as part of their synoptic assessment. Investigations are ideal ways for you to display – and for boards to examine – your synoptic skills. You saw in Chapter 2 that synopsis can be thought of as the use of skills in contexts which bring together different areas of the subject. Practical work is an excellent way of doing this. It is at the heart of what many professional physicists do for their work. Success at practical physics is a mark of your competence as someone who may be about to enter the profession.

Choosing what to do

Before you can get very far with any piece of physical investigation you need to choose a topic. The degree of choice you will have here depends on a number of factors: the way practical investigations are organised in your school; the timetable laid down for the work; and the rules of the examination. Your success can depend heavily on an appropriate choice of an investigation topic. So where do you begin?

The first question you should ask concerns the way that your work will be assessed when you have handed in the final report of the investigation. It is important that you ensure that you can collect all the marks that are available. So you should take care that you do not shut yourself off from whole areas of marks by failing to understand the marking criteria that will be used in judging your work.

Here are some of the marking criteria for one of the boards that uses practical investigation.

Part of the marking criteria for the WJEC specification

Each of the 15 criteria below are marked on a scale of 0 (worst) to 3 (best) with 45 marks available for the whole investigation.

The marks are allocated (for each criterion) as follows:

0   Only awarded if a candidate shows no achievement in the skill tested.

1   Some attempt is made to satisfy the criterion.

2 There is clear evidence that the candidate has demonstrated skill in the criterion.

3 The candidate has shown a high level of skill and has communicated this effectively within the coursework.

Even if you are not taking WJEC examinations, this set of criteria makes a good starting point for your thoughts on what constitutes a good and complete investigation.

A: Planning the investigation

Candidates should have:

1 considered methods appropriate to an investigation at a full A-level standard;

2 correctly defined the variables likely to feature within the investigation;

3 planned an effective strategy for answering the problem.

B: Implementing the experimental work

Candidates should have:

4 adopted effective techniques and safe working procedures;

5 selected appropriate instruments and procedures to give the accuracy required;

6 demonstrated that they have followed their outline plan and built in modifications in the light of results if appropriate.

C: Recording and collecting data

Candidates should have:

7 made and recorded in table format sufficient relevant observations and measurements to the appropriate degree of precision;

8 displayed the units, instrument resolution and zero errors;

9 evidenced ability to present graphs accurately in terms of labelling scales, units and errors bars.

D: Analysis and evaluation

Candidates should have:

10 used data appropriately to analyse the identified problems;

11 expressed an awareness of the limitations for experimental measurements by quoting any result to an appropriate number of significant figures;

12 critically analysed the experimental procedure and the reliability of the data collected; evaluated the techniques used and suggested improvements where appropriate.

E: Communication

Candidates should have:

13 presented a valid conclusion, which provides an appropriate response to the identified procedure;

14 presented a well-structured and concise report employing a range of vocabulary and modes of presentation;

15 demonstrated appropriate IT skills.

You may be allowed to see examples of work submitted at your centre in previous examination years. You should take full advantage of this if it is possible. Look at reports with high marks, noting the particular qualities of the work. Also you should try to work out why the students with low marks scored badly. There is, however, one important thing to remember when you look at other people's work: you only see the end result, the final report. Behind this was a fortnight or so of practical work and development that does not always appear in the written accounts. That having been said, a good report usually reflects the way in which the student refined and improved both apparatus and ideas as the work progressed.

You will need to discuss your plans at an early stage with your teachers. Try to keep an open mind for as long as possible. Your teachers not only know the availability and the quality of apparatus which they can supply to you, but they will also know your limitations and they will try to prevent you from choosing a project that is too difficult. Listen carefully to what they have to say and modify your plans in the light of their comments.

Try to let the project grow out of your personal interests. If you play a musical instrument, why not investigate the way the sound is produced. A string player might look at the variation of note frequency with string tension or some other parameter of the string. If you play a sport, consider some aspect of the mechanics of the game: the 'sweet spot' on the racquet in tennis or badminton for example. Devise an investigation of why a cricket bat sometimes jars the wrists or about the power delivered to a pedal cycle and so on. Photographers, artists, stage technicians can all find some aspect of their art or craft to investigate. Just think creatively.

If you really are stuck for a topic, the websites of some awarding bodies (the WJEC is an example) contain investigation titles that you could consider.

Try out some preliminary experiments before the time set for the investigation proper. You will almost certainly get some credit for this provided that you retain the preliminary data and incorporate them into your report. It may not be possible for you to be allowed advance access to the equipment you require, but if you can, try to see whether your ideas are going to work. Most A-level investigations can be set up relatively quickly during a lunch time, and this, at least, should enable you to get a feel for the results you are likely to get and the size of the effects you hope to measure. It will also show if there are any major defects in your apparatus design that will need to be corrected before the investigation proper begins.

There may be some aspects of your work that can researched through existing books or through the Internet. See what others have done by all means. But if you do use anyone else's work (whether their words or their data), you must give them credit in your report – and it is sensible to gain permission from your teacher to use other people's work before you begin.

The worst way to investigate is to come to it cold, with no ideas other than the basic title and with no thoughts or preliminary ideas of where the work is likely to go.

The investigation needs to have a sense of flow and development. Good investigators at any level are rarely doing exactly the same experiment at the end of the project as at the beginning. Begin with simple ideas and relatively simple experiments and develop these as your ideas grow or as some of your earlier ideas prove to be unsuccessful.

The very best topics to choose will enable you to collect numerical data and will be problems that are well defined. *What happens when I heat golden syrup?* may not be as good as *How do the flow properties of golden syrup change with temperature.*

When you think you have found the perfect investigation title, ask yourself some difficult questions about the topic:

★ Is the physics of an A-level standard?

★ Can plenty of measurements be taken in the time available and still leave time for thinking about and analysing the physics?

★ Will the preliminary work of the first few hours permit development into better and more refined experiments?

★ Will you be able to draw conclusions from your experiments?

★ Are you going to be as interested in the experiment and its results on the last day as on the first?

If the answers to all these are 'yes', then go ahead, it is probably the perfect project for you.

To sum up:

Choose an investigation in which you can:

● **Use your understanding of physics to plan out an investigation and choose appropriate experiments for it.**

● **Be safe, and demonstrate that your working is safe.**

● **Show initiative and independence in doing the experiments and interpreting the results that they give.**

● **Make and record the observations appropriately with appropriate instruments, used correctly.**

- Analyse the results and then evaluate what you have done to check how reliable the results are.

- Communicate the results and conclusions well to an A-level standard using as many of the techniques that you have been taught as you can (graphs, tables, and so on).

Your choice should enable you to demonstrate synoptic skills both in the work that you do and in the way you write the work up.

Recording the investigation

Here are some guidelines that you might wish to consider about the day-to-day running of your work. This advice will apply to many physics projects both at A-level and beyond.

It is imperative that you keep an accurate, daily record of what you do in the laboratory. The form you choose is a personal matter, but many professional scientists keep their records in a hard-backed notebook so that there is no danger of stray pieces of paper falling out and being lost. You may wish to keep loose-leaf notes in a file, which is fine but make sure that you do not lose any. If your records are kept in an electronic form in a computer, back everything up regularly – a daily backup is a good routine to have.

Every day spend some time reviewing the work you have done: work out any results, draw the graphs of any data you have taken during the day. Think about where the work will go next time and about what you hope to achieve in the laboratory – and then write down your aims and targets in your notebook. This will allow you to plan your work ahead. Make the most of your laboratory access and do your planning and thinking at home.

Keep asking yourself the question: where is the work going? Can you explain the results you have so far, do they make physical sense, can you present the results in other ways that will enhance their meaning?

Writing-up

You will almost certainly be required to produce a formal report of your investigation. Your teachers will guide you about the form that this should take, but here are some ideas that you may wish to consider.

Begin by writing a brief summary of the whole project, say what your investigation aim was at the outset of the work, and say to what extent you achieved these aims.

Discuss early on in the report the underlying science on which your work is based. You will need to give a reference to any data that you are using from someone else. There is a particular way to do this that is agreed by all scientists, the idea being that any one else can find the particular article that you are referring to. You can find details of this system in the written-report section of Chapter 4 (page 50).

It is in your interests to mention in your report any difficulties that you had or any blind alleys you followed for a while. An A-level practical project is as much about your ability to carry the work through and to develop it, as it is about your ability to invent new science. In the short time allotted to your practical project, you are unlikely to make discoveries leading to a Nobel prize. So, put down everything, using the daily record of your work to remind you of the problems as well as the triumphs. Give details of your preliminary results including how you did the tests, and graphs or tables showing the results themselves.

You may be required to give details of safety arrangements you took. Certainly, if there were any corrosive chemicals or electrical experiments your teachers may have insisted on a risk assessment. Put these in the report too.

Include details of all the apparatus you used. (A diagram is usually sufficient, or you may wish to use a photograph. A digital camera, if you can gain access to one, makes it particularly easy to add pictorial material to a word-processed report.)

Graphs and diagrams should be added to the text at the place where they are discussed first, rather than collected together at the end. On the other hand, large tables of results are best kept to the end and placed in an appendix.

The analysis comes after your account of the experiment, followed by the conclusions you draw. Remember your training at GCSE and the importance then of evaluating your work. It is important to comment on the results themselves, drawing attention to anomalies in the results, although you should have tried hard to iron these out. Remember to discuss the extent to which your results support your conclusions and try to account for reasons why this support may be less than you would have liked. Be critical about your work; no one does perfect science. One of the marks of the good scientist is the ability to be self-critical.

Finally, here is a checklist of items that ought to find a way into your report. Only omit these with a good reason.

A report should contain:

- **Statement of aims and a brief summary of the work as a whole**
- **Outline of the physics behind the project**
- **Details of preliminary work**
- **Account of the work done, possibly in a diary format rather than apparatus/method/results/conclusion (don't forget safety details)**
- **Results and analysis of the results, tied together with an account of the physics**
- **Conclusion which presents the findings together with an evaluation of the investigation as a whole.**

Practical examinations

Practical examination papers are set by three awarding bodies. Like investigations, practical tests are well suited to a synoptic assessment as they allow the examiners to draw widely from the entire syllabus and to test these areas in a context that is at the core of the practical physicist's work. The key is to remember that practical examinations in physics are *not* tests of your practical ability. They are not, in other words, tests of how well you can wire up an electrical circuit or measure the time period of a swinging pendulum. Instead the practical papers test your ability to think within a practical context, they test if you can take data by yourself, and then analyse them to reach a conclusion – and can you do it against the clock?

Before the examination day

As always, careful preparation is the key. The practical questions can come from virtually any area of the syllabus, although for pragmatic reasons there tends to be an emphasis on questions that will not burden examination centres too heavily. So, mechanics experiments and electrical experiments tend to be common, with radioactivity experiments or those that involve expensive pieces of equipment, such as oscilloscopes, quite rare. So, know the specification material just as well for a practical test as for the theory papers.

Additionally, make sure that your practical analysis skills are up to scratch. Be ready to make decisions about the best graphs to plot (Chapter 5). Draw the graphs and carry out calculations that arise from them accurately and quickly (Chapter 6). Be prepared to write about the experiments and to suggest improvements to both the apparatus and the measurement technique.

One board (AQA specification B) issues a briefing sheet for one of the two practical papers set at the end of the course. You should read this carefully and try to look up the physics of the practical context in the days before the practical test itself.

Whatever board you are taking, ensure that you know what the shape of the practical paper will be. How long the paper will last, how many questions there will be and what the mark weighting between questions is.

Other questions you should have about the tests include, what have previous papers looked like and how were the marks allocated on the mark scheme? If you can find out what examiners have said in the past about candidates' graph and analysis skills, then you will know the mistakes to avoid.

On the day
Do what the examiner asks

As ever, it is crucial that you read the questions carefully. Not only do you have to grasp the physics of the questions that are being asked, but you also need to understand completely what you have to do in the practical part of the examination. You must do the experiment that is asked. If you do not, then it will be difficult for the examiner to award marks to you. However good your data collection and your write-up, if the work you have done does not match the mark scheme then you will gain very little credit.

The bifilar pendulum and electrical circuits are two examples of the way that things can inadvertently go wrong.

The bifilar pendulum is rarely taught at A-level. However, the theory is reasonably straightforward and similar in many ways to the simple pendulum with which you may be familiar. Also, the bifilar suspension itself is easy to set up and uses common inexpensive pieces of apparatus. This makes it an obvious possibility for an examination question.

The pendulum can be constructed from a metre rule and two lengths of string (Figure 7.1). All fairly simple so far. Unfortunately, having set the system up, there are three different modes in which it can be made to swing. The time period and the mathematics depend on which mode is in use – use the wrong one and the experiment will not work as the examiner expects. The examination question will be carefully written so that the swing mode is described unambiguously, there may even be a diagram to confirm matters. But careless candidates working under pressure will use the wrong swing and perform the wrong experiment.

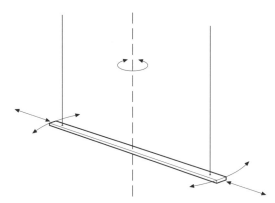

Figure 7.1 A bifilar pendulum and its swing modes

Another danger area is setting up electrical circuits. Do make sure that you have used the specified components and connected them up correctly to match the circuit diagram.

One tip is to arrange the components on the laboratory bench exactly as shown on the diagram (Figure 7.2). This is an obvious point but a simple one to carry out and to check.

Another way in which you can make electrical circuits more straightforward for yourself under examination conditions is to have a routine sequence of setting up (Figure 7.3). The order is: series parts first, parallel afterwards. Again the diagram will make this clear.

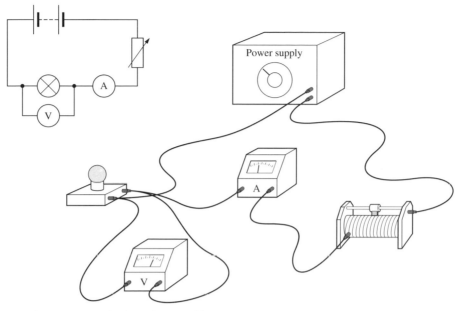

Figure 7.2 The circuit on the bench resembles the circuit diagram

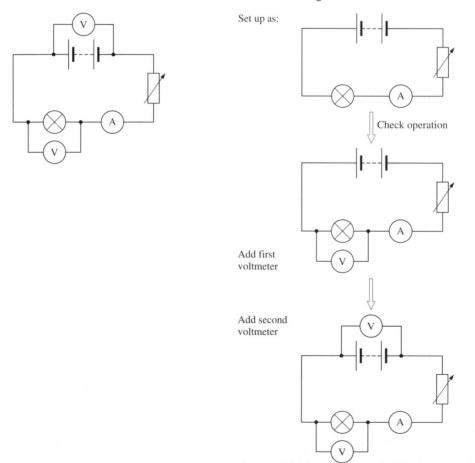

Set up as:

Check operation

Add first
voltmeter

Add second
voltmeter

Figure 7.3 Series first, parallel last

If in doubt, ask

Whatever the context, all practical examinations allow you to ask for help and advice if the experiment does not appear to work. **Ask for this advice if you are in any doubt about the apparatus.** First of all, apparatus does break down during a practical examination. If the apparatus is faulty, then you will be given a set that works properly and also some extra time.

Alternatively, it may be that you made a mistake and the apparatus is not faulty at all. It is still worth asking for help because few examiners penalise by more than two or three marks for help from a teacher. You have to balance this loss of a few marks against the loss of many marks if you continue to struggle on unsuccessfully for a long time, attempting to set up a difficult electrical circuit or other piece of apparatus.

On the whole, it is very much better to accept the loss of the few marks in the hope that you can go on to collect a good set of data for your analysis. No data, no analysis, no marks!

Getting the timing right

One of the pieces of research you need to do before the examination day is to find out the structure of the paper. You need to do this so that you know how much time is allocated to each question. You also need to know how your school physics department will arrange the examination. It is likely that the teachers will group you and your fellow students so that less apparatus has to be prepared.

Take for instance an examination paper of 2 hours with three practical questions, two timed at 30 minutes each with the third lasting 60 minutes. Half of your group of students will be set the 60 minute question first with the remainder divided into two further groups with 30 minutes per experiment. Every 30 minutes the group with the short questions will rotate on to the other question. Half way through the examination, the long-question students will move to the short questions and *vice-versa*. Under these arrangements, you will only receive the correct time with the apparatus.

The moral is: keep your eyes on the clock. The essential point is to make sure that you have completed the data collection in the time available. You may well be able to claw back some time to complete the analytical work during one of the other questions. But if you have not even collected the data in the first place then this will not be possible.

This is particularly important if the examination has five or six very short questions. It is hard to write questions that are exactly the same time length as each other. Use the time left over at the end of the shorter questions to finish off earlier ones or to look ahead at questions to come. Above all, do not sit staring into space just because you have finished one of the short questions ahead of the time. Keep focused!

Getting the data right

Another very important detail is to take the correct number of data points. Sometimes you are told how many observations to make, you must do this correctly. The consequences of getting this wrong can be surprisingly far-reaching. You will lose marks for disobeying the instructions, you will lose marks for having the wrong number of data points on the graph, you will lose marks for deriving quantities from the wrong number of points and finally your answers may well be less accurate than required.

If the examination paper does not suggest a number of observations, you will have to be more cautious and rely on your experience. Think about a minimum of six or seven, with even more for an experiment with data that are not very reproducible.

This brings in the question of repeating individual data points as a check: do so wherever possible. Look into repeating each measurement once or twice and then take the average of the two or three values. Again, this requires judgement, if the measurement can be taken quickly and there is a lot of spread in the values, take more.

You may have to choose the range of data values too. Take a large range of values, the whole range that the apparatus will allow. Normally, the independent variable values should be distributed evenly right across this range. The only exception would be where a reciprocal or a semi-logarithmic graph is to be plotted. The effect of taking data at equal steps and then working out the reciprocal or the logs means that the data become squeezed into one corner of the graph, try to anticipate this and spread your data accordingly.

Having taken the correct number of data points and repeated them where appropriate, you must use them. Do not omit data points from your graph without a very good reason. Even then, make it clear to the examiner what is going on. The very best thing is to repeat the measurement of any data point that seems to be anomalous in some way. Do not try to get out of these problems by subterfuge. The examiner will spot straight away that you have tweaked your graph axes so that a 'difficult' point falls outside the range of the graph grid. You will lose marks both for a poor scale and for omitting points! It is simply not worth it.

As far as possible try to obey the suggestions made in the graphing skills sections of Chapters 5 and 6.

Planning, evaluation and design questions

So far this section has concentrated on the type of examination question where you are told what to do. Practical tests can be rather broader than this and nowadays there can be aspects of planning, evaluation and design in the questions. You will have been introduced to these skills at GCSE, so A2 builds on the same ideas in the context of the more advanced material you have been learning since year 11.

Planning and designing

Planning is more than simply describing a set of apparatus and a method for the experiment. It also involves making predictions based on theory and therefore suggesting the analysis that is needed and the outcome expected. In the examination you may well be given a rather crude experiment to perform quickly and then asked to suggest improvements. You may not necessarily have to carry out this designed experiment.

As usual try to have a set routine so that under pressure all your skills do not desert you.

Planning an experiment

- What are the variables?
- Which variables are you choosing (or are you told) to vary?
- What is your prediction and how does it follow from physics?
- What graph will you plot to illustrate this prediction and what shape will it be if the prediction is correct?
- What measurements will you be able to derive from your data or the graph?
- How will you vary the independent variable and measure it?
- How will you measure the dependent variable?
- Is the experiment safe? How can you improve the safety?
- How will you display and manipulate the data to produce the graphs you require?
- How will you derive data from the graph to enhance your prediction?

These boil down to:

- *Predictions* and graphs that lead from them;
- Making *measurements* safely and displaying them;
- Data *manipulation* and verification of prediction.

Evaluating

You may be asked to carry out an evaluation of the experiment you have carried out or to suggest (design) improvements to the technique and apparatus or to propose further experiments.

Here is a check list.

Evaluating an experiment

- To what extent can you trust your data? Do they have large uncertainties?
- In the light of the answers to these two questions: to what extent have you verified your prediction?

● Are there any anomalous results? To what do you attribute these, to something in the physics or to something in the experiment?

● What are the weaknesses in your experiment? How can you eliminate them?

● How can you build and improve on the strengths in your experiment?

● Have you seen or performed an experiment during your course that does the job better in some way? Describe it.

● Would a choice of different instruments make the experiment easier or quicker or more precise?

● Would data logging help in any way?

Example practical question: This question is about the power delivered by a power supply to a load resistor.

a)   Connect up the circuit shown in Figure 7.4.

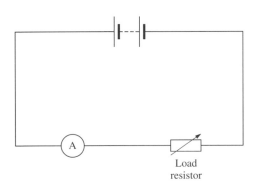

Figure 7.4

b)   In this experiment, you will be measuring the current for six values of load resistor. You will use these data to find two further derived values for each pair of current–resistance measurements.
Draw up a table suitable for recording all these values.

c)   Measure the current in this circuit and record the value of the load resistance.

d)   Repeat this measurement for the five other load resistors.

Table 7.1 gives the experimental data if you are answering this without access to equipment.

Reading no	1	2	3	4	5	6
Load resistance/Ω	100	150	270	330	470	580
Current/mA	14.6	13.0	10.3	9.4	7.7	6.7

Table 7.1

e) Use your data to calculate the power dissipated in each resistor and include these derived values in your table.

f) Plot a graph of power dissipated in the resistor against load resistance.

g) Use this graph to estimate the resistance for which the power delivered to the resistor is a maximum.

h) Show that the emf ε_0 of a cell with internal resistance r is related to the current I in the cell and the load resistance R by:

$$\varepsilon_0/I = R + r$$

i) Add the derived value $1/I$ to your table of results. Plot a suitable graph that will enable you to determine the emf and internal resistance of the cell. Go on to calculate these values.

j) Comment on your answers to parts (g) and (i).

Suggested answers and hints

a) Make sure that the circuit is correct. Check that the variable resistor is set correctly.

b) There will eventually be six data points. The question tells you that for each data point there will be readings of current and resistance together with two additional values derived from your data.

c) Read the ammeter correctly, especially if it is an analogue type.

d) Re-set the variable resistor correctly and note its value with care.

e) (power) = (current)^2 × (resistance).

Load resistance/Ω	100	150	270	330	470	580
Current/mA		14.6	13	10.3	9.4	7.7	6.7
Power/mW		21.316	25.35	28.6443	29.1588	27.8663	26.0362

Table 7.2

f) Figure 7.5 shows graph of power against load resistance.

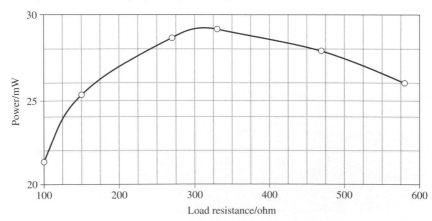

Figure 7.5

g) The maximum power occurs when the load resistance is 310 Ω.

h) ε_0 = potential difference across load resistor + p.d. 'lost' in cell

$$\varepsilon_0 = IR + Ir$$

$$\text{so} \quad \varepsilon_0/I = R + r$$

i)

Load resistance/Ω	100	150	270	330	470	580
Current/mA	14.6	13	10.3	9.4	7.7	6.7
Power/mW	21.316	25.35	28.6443	29.1588	27.8663	26.0362
1/Current/mA^{-1}	0.068493	0.076923	0.097087	0.106383	0.12987	0.149254

Table 7.3

i) $1/I - R/\varepsilon_0 + r/\varepsilon_0$ compare with $y - mx + c$.

Plot $1/I$ against R, gradient is $1/\varepsilon_0$, intercept on y-axis is r/ε_0 (Figure 7.6).

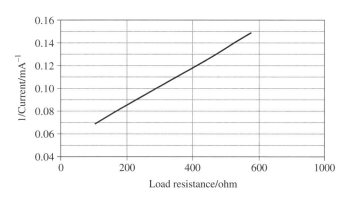

Figure 7.6

Emf = 6.0 V; internal resistance = 310 Ω.

j) The maximum power is delivered to load resistance when its value is equal to the internal resistance of power supply.

Question 1 – speed of water waves: This question concerns the speed of a water wave and its variation with depth.

a) Figure 7.7 shows a shallow trough 0.45 m long that can contain water to a depth of about 0.1 m.

Pour water into the trough until it reaches a depth of 0.01 m. Lift one end of the trough by about 10 mm and lower it again; this should make a pulse of water that will travel down the trough, reflect at the far end and then return.

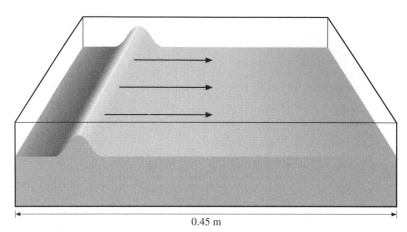

0.45 m

Figure 7.7 A wave travelling down the tank

b) Take measurements to determine the speed of the water wave along the trough for this depth of 0.01 m.

   (i) State the measurements you took and explain how you arrive at your answer for the speed.

   (ii) Tabulate your measurements preparing a table that will accommodate five further sets of measurements.

Table 7.4 gives the experimental data if you are answering this without access to equipment. The student who took these measurements timed how long it took for a wave to make five two-way crossings of the tank.

Depth/m	Time for 5 crossings/s		
0.01	14.5	14.3	14.3
0.02	10	10.3	10.2
0.04	7.2	7.2	7.3
0.06	5.7	5.9	5.9
0.08	5	5.1	5.1
0.10	4.5	4.6	4.6

Table 7.4

c) Take further sets of measurements of wave speed for five further depths of water between 0.020 m and 0.10 m.

d) (i) Estimate the absolute uncertainty in the measurements that you made.
   (ii) Use these estimates to calculate the uncertainty in the speed of the wave.

e) There is a suggestion that the speed of water waves varies with water depth according to the relationship:

$$v = (kd)^n$$

where k and n are constants.

Plot a suitable graph of your results and use it to evaluate n and k.

f) State the units of k.

Question 2 – Emf induced in magnetic coils: This question asks you to investigate how the separation of two coils affects the voltage induced in one of them.

a) Set up the circuit shown in the Figure 7.8.

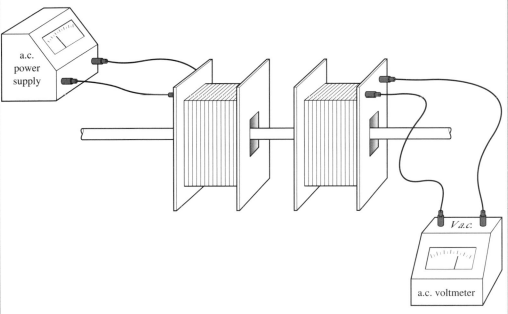

Figure 7.8

Set the coils so that they are touching in the middle of the rod. Measure the distance between the centres of the coils and the corresponding value of the induced voltage on the secondary coil.

b) You will be making six observations of data points in this experiment. Draw up a suitable table for the display of your results.

c) Record the values of separation and induced voltage for the case when the coils touch.

d) Repeat the experiment for five further separations up to a coil separation of 6d, where d is the distance between coil centres when coils are just touching.

Table 7.5 gives the experimental data if you are answering this without access to equipment.

Reading	1	2	3	4	5	6
Separation	d	2d	3d	4d	5d	6d
Induced emf/V	5.2	3.4	2.2	1.5	0.9	0.6

Table 7.5

e) There are two suggestions for the relationship between induced voltage and coil separation.

Suggestion 1. Induced voltage is inversely proportional to coil separation ($V \propto 1/d$).

Suggestion 2. Induced voltage varies exponentially with coil separation ($V \propto \exp(-kd)$).

Draw two suitable graphs which will demonstrate the correctness or otherwise of these suggestions. State which suggestion is most likely to be correct.

f) Estimate the induced voltage that you would obtain if the coils were wound on top of each other.

Question 3 – Stressed wooden beam: This question concerns the behaviour of a stressed balsa beam.

a) Figure 7.9 shows a wooden metre rule beam clamped to a laboratory bench with a mass hanger suspended from the end of the rule.

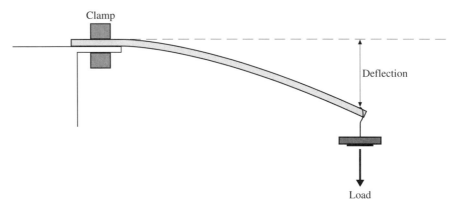

Figure 7.9

Devise an experiment to measure the deflection of the end of the beam as the load hung on the beam varies. The metre rule will support a mass of 0.30 kg before breaking.

b) Record your observations in the space below.

Table 7.6 gives the experimental data if you are answering this without access to equipment.

Mass on beam/g	50	100	150	200	250	300	350	400	450	500
Deflection/mm	3	5	8	11	14	17	20	22	25	28

Table 7.6

c) Give an account of the experiment you devised.

d) Plot a graph of load against deflection.

e) Robert Hooke stated, in a famous law, that the force exerted on a spring is directly proportional to the extension of the spring. Discuss the extent to which this beam acts as a spring that obeys Hookes' law.

f) Theory predicts that for a beam bent in this way:

$$\text{Deflection } y = \frac{4Fl^3}{Ebh^3}$$

where F is the force applied to the end of the beam, l is the length of the beam, E is the Young modulus, b is the width of the beam and h is the thickness.

Table 7.7 gives the results obtained by a student. Use these data together with your graph in order to calculate the value of E including an estimate of the error in the measurement.

b/cm	h/mm	l/cm
2.54 ± 0.08	4.8 ± 0.1	45 ± 1

Table 7.7

CHAPTER EIGHT

PRACTISING SYNOPTIC SKILLS

This chapter contains two questions: one a comprehension exercise, the other a data-analysis test, both with suggested solutions. Chapter 9 contains more comprehension and data-analysis questions for practice. There are answers for all these questions in Chapter 10.

There is a description of a strategy for answering trial questions in Chapter 3. As a reminder, if you cannot remember the physics after reading a question thoroughly, consult your notes and text books carefully, paying particular attention to any worked examples that are similar to the question, expand your notes at this stage if you think it necessary, and then – having closed your notes and books – return to the original question. Try not to answer the practice question with your notes and books in front of you. You need to develop confidence that you can answer the questions unaided.

Comprehension questions

Electricity in the kitchen (time allowed – 1 hour)

Microwave ovens

Labour-saving and time-saving devices powered by electricity continue to improve the lot of both the domestic and the professional cook.

The microwave oven is now a common sight in many kitchens. The oven can be used for the rapid heating of pre-prepared and fresh food. It produces microwave radiation that can be absorbed readily by water. Water molecules have charge distributions that are asymmetric and a resonant molecular oscillation can be stimulated if the microwave frequency is correct. The frequency of the microwaves used in the oven is 2.45 GHz.

The process begins in the microwave oven when electrons are accelerated in a device called a magnetron. The waves are injected into the interior of the oven where they set up a standing-wave pattern. This pattern can be very complex but, even so, may well have regions where there is little or no excitation of water molecules. In order to combat this problem, manufacturers often fit a turntable in the oven or advise users to rotate the food several times during the cooking process.

➤

The microwaves can only penetrate a short distance into the food. Typically, over a penetration depth of 3 mm the intensity of the wave will fall to 40% of its original value. Thicker food can, however, be cooked through processes other than microwave penetration.

Common uses for the microwave in the home include the rapid thawing of frozen foods and its subsequent reheating (often in its original storage container which may suffer little direct heating from the microwaves).

Induction hobs

Alternating-current effects are used in another modern cooking context, in induction hobs on cooking stoves. These have the advantage that current is induced in the pans but the stove itself never becomes hot.

Inside the stove, underneath areas marked on a glass top, are coils of wire. An alternating current is fed to these coils; the frequency of the current is much higher than the mains and can exceed 18 kHz. A large alternating magnetic field is produced in and around the coil and this induces an emf and therefore an induced current inside the cooking vessel.

Perhaps, surprisingly, copper is not the best material for the cooking pots. Iron and steel have the advantages of larger resistivities than copper and a skin effect that confines the induced current to the layer of metal close to the surface. This further increases the effective electrical resistance.

A similar principle can be used to light the gas in a more conventional hob. Rather than use the trusty match, modern stoves have an electronic device that will supply a spark quickly to avoid the build-up of dangerous unburnt gas. A small potential difference is used to drive an oscillator circuit with an output of 25 V pulses connected to a coil. This coil is the primary for a transformer that steps the voltage up to pulses of about 300 V which are used to charge a 1.5 µF capacitor.

The oscillator is running at a frequency of about 5 kHz and after about 1000 pulses the capacitor will be fully charged. It is then discharged rapidly through the primary of another transformer that produces 15 kV or so at a spark gap near the gas supply to an individual burner. The energy of each spark is about 1 mJ. This value is independent of the state of the small 1.5 V battery used to power the device and this ensures that the quality of the ignition does not vary.

Pressure cookers

Do not think, however, that all modern cooking relies on recent and high-tech innovations, many domestic cooks still use the well-tried and familiar pressure cooker. This device consists of a sealed container into which the food to be cooked and some water are placed. The water is raised, under pressure, to a temperature above 100°C and this means that the food cooks more quickly than it would in a pan of boiling water.

➤

One way to control the pressure is through the use of a tapered pin seated in a tube leading to the pressure vessel. The pin is held in place with a spring. When the pressure exerted downwards by the outside atmosphere and the spring is less than the total pressure in the cooking pot, then the tapered pin rises and steam is released from the vessel until the pressure drops to the control value. This presents an effective solution in the modern kitchen after more than a century of use.

c, speed of electromagnetic waves in a vacuum $= 3 \times 10^8$ m s^{-1}.

1 a) Show that the wavelength of the microwaves used in a domestic microwave is about 120 mm.

b) Figure 8.1 shows the interior of a typical microwave oven. Microwave energy is injected along the horizontal dotted line shown. Consider only waves travelling along this line. Draw a diagram showing the simplest standing wave pattern along this line.
Mark **one** position where food is unlikely to cook well.

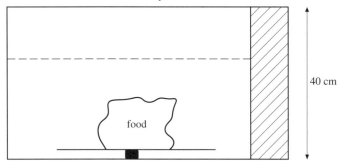

Figure 8.1

c) Suggest why the passage states that 'the pattern can be very complex' and go on to state why some manufacturers 'often fit a turntable in the oven' which constantly changes the position of the food during cooking.

d) Manufacturers recommend that no metallic containers are used in microwave ovens. Explain why.

e) (i) Use information in the passage to draw a graph of microwave intensity against depth in the food.
   (ii) Use your graph to estimate the depth at which 90% of the microwaves have been absorbed by the food.
   (iii) Explain why food much thicker than this can be cooked thoroughly.

f) Estimate the minimum time it will take a 750 W microwave oven to thaw and then heat 0.3 kg of frozen soup if the soup comes straight from the freezer at −18°C and is to be heated to 90°C. Assume that the soup is entirely made of water.

Energy to heat 1.00 kg of water through 1 K = 4200 J

Energy to heat 1.00 kg of ice through 1 K = 2200 J

Energy to change 1.00 kg of ice at 0°C to water at 0°C = 334 000 J

2 a) Figure 8.2 shows a saucepan being placed on one of the induction hob coils.

Figure 8.2

Draw a diagram of the coil and add to it a sketch of the magnetic field when the current in the coil is a maximum.

b) Draw a separate sketch of the saucepan and on it show the paths of the currents induced:

(i) in the base of the saucepan
(ii) in the walls of the saucepan.

c) The cooking vessel and the high-frequency coil can be considered together as a transformer (Figure 8.3).

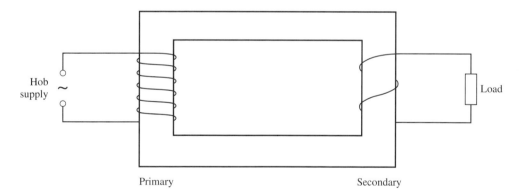

Primary   Secondary

Figure 8.3

Estimate the size of the current that will flow in the cooking pot if energy is being transformed at a rate of 2 kW. Assume that the effective resistance of the cooking pot is 0.2 Ω.

d) An induction hob is rated at 2800 W. The frozen soup of Question 1(f) is contained in a copper saucepan of mass 0.90 kg. 15% of the heat energy generated is lost to the surroundings. Estimate the time to heat the soup on the induction hob.

Energy required to heat 1.00 kg of copper through 1 K = 300 J

e) Explain why a large frequency must be used for the current to the induction coil.

3 **a)** Use the information in the passage about the gas lighter to show that the main capacitor stores about 70 mJ before discharge occurs.

   b) The capacitor is charged using a 1.5 V cell. Estimate the maximum frequency of sparking.

4 Explain how raising the pressure in the pressure cooker above atmospheric pressure enables the food to cook more quickly.

5 The rate of cooking in the pressure cooker is related to the temperature inside by the equation:

Rate of cooking $= \text{constant} \times e^{-E_0/kT}$

Where E_0 is a constant with the value 2.3×10^{-19} J for cooking potatoes, k is Boltzmann's constant (1.38×10^{-23} J K^{-1}), and T is the absolute temperature.

It takes 20 minutes to boil potatoes in a pan of boiling water. Calculate the temperature required inside the pressure cooker for the cooking time to become 5 minutes.

6 The passage describes one way of controlling the pressure in the cooker. Figure 8.4 shows how this might be achieved in practice.

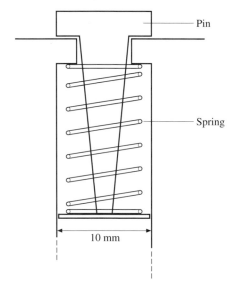

Figure 8.4

The safety limit for a particular cooker is 50 000 Pa above atmospheric pressure and at this temperature the pressure will lift the pin away from its seating against the force in the spring. Calculate the force that the spring will need to exert.

Suggested answers

1 **a)** $c = f\lambda$; $\lambda = \dfrac{3 \times 10^8}{245 \times 10^9} = 122$ mm

b)

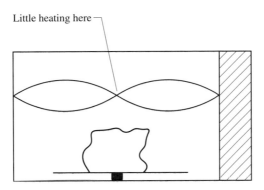

Little heating here

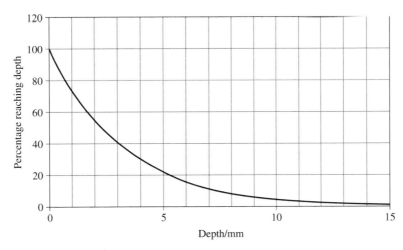

Figure 8.5 Penetration of microwaves

c) Reflections at walls and objects in the oven lead to complex standing wave patterns where waves add and subtract; the turntable sweeps food through 'hot' and 'cold' spots ensuring even heating.

d) An electromagnetic wave has both electric and magnetic components. The changing magnetic flux leads to induced emfs and, if there is a complete low-resistance electrical circuit, induced currents, which are often very large. These currents lead to sparking and damage to oven and cookware.

e) (i) 3 mm leads to 40% of intensity; 6 mm to 40% of 40%, i.e. 16%, 9 mm to 40% of this and so on.
   (ii) Read-off from the graph shows that about 8 mm of food absorb 90%.
   (iii) The principal mechanism is thermal conduction through solid food.

f) 300 s (heat ice, 11.8 kJ; melt ice 0.1 MJ; heat water, 0.113 MJ; divide by 750 W for time).

2 a) Figure 8.6 and **b)** Figure 8.7

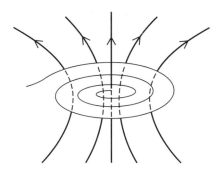

Figure 8.6

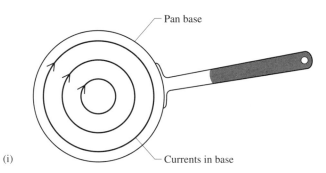

Pan base

(i)

Currents in base

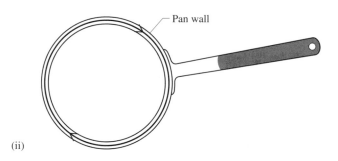

Pan wall

(ii)

Figure 8.7

c) $I^2R = 2000$; $I = \sqrt{(2000 \div 0.2)} = 100$ A

d) Energy to heat soup = 225 480 J; energy to heat saucepan = $108 \times 0.9 \times 330 =$ 29 160 J; total energy = 254 640 J, so time = $254\,640 \times 115/(2800 \times 100) = 105$ s.

e) Induced emfs and hence induced currents are proportional to the rate of change of flux. High frequencies lead to rapid rates of change and hence large currents.

3 a) energy stored = $V^2/2C = \dfrac{300^2}{2 \times 1.5 \times 10^{-6}} = 67.5$ mJ.

b) 150 pulses required at 5000 s^{-1}, so 30 Hz is the frequency.

4 Increased pressure leads to increase in boiling point, so steam in vessel has higher temperature (Boltzmann factor increased).

5 385 K.

6 $p = 50\,000 = F/A$; $A = \pi \times$ (radius of pin)^2; $F = 3.9$ N.

Data analysis questions

Cavity wall insulation (time allowed – 45 minutes)

Some modern British homes are designed with cavity walls. Figure 8.8 shows the cross-section of this design. In essence, there are three layers to the outside surface of the dwelling: an aerated concrete block, an air cavity, and as the outside layer, a layer of house bricks. This simple model ignores the presence of decorative materials on the inside surface such as plaster and wallpaper.

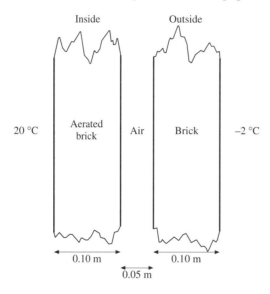

Figure 8.8

Some tradesmen offer a service in which they fill the cavity itself with a foam product which, it is suggested, will reduce heating costs by reducing the flow of heat through the wall. This product is often known as 'cavity fill'.

The equation that describes the flow of heat through a material is:

$$\Delta T = \varphi R$$

where ΔT is the temperature difference across the material, φ is the rate of transfer of thermal energy through unit area and R is the thermal resistance. This equation leads to a definition of thermal resistance:

$$R = \frac{l}{kA}$$

Suggested answers

1 a) $X = RA; = \frac{l}{k} = \frac{0.1}{0.9} = 0.11$ m² K W⁻¹.

b) $\Delta T = \varphi R; \varphi = \frac{\Delta T}{R} = \frac{1}{0.11} = 9.0$ W m⁻².

c) (i) $X_{concrete} = \frac{0.1}{0.4} = 0.25$ m² K W⁻¹. (ii) $X_{brick} = \frac{0.05}{0.28} = 0.18$ m² K W⁻¹.

2 a) $R = \frac{V}{I}$

b) $\rho = \frac{RA}{l}$

c) pd ΔT, current φ, thermal resistance R.

d) $R_{total} = R_1 + R_2 + R_3$.

3 a) Add the individual thermal resistance coefficients.

b) $0.11 + 0.25 + 0.18 = 0.54$ m² K W⁻¹.

4 a) $\Delta T = \varphi R; \varphi = \frac{22}{0.54} = 40.7$ W m⁻².

b) $\Delta T = 40.7 \times 0.25 = 10.2$ K.

c) $\Delta T = 40.7 \times 0.15 = 7.3$ K.

d)

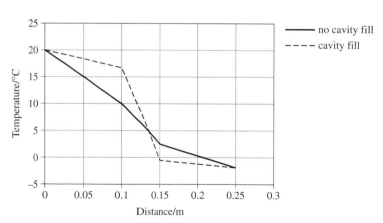

Figure 8.9

5 a) 13.7 W m⁻².

b) See Figure 8.9.

6 The thin air layer creates a large temperature gradient near to wall and this effectively insulates the surface even more. φ will decrease.

7 a) Straight line from (0.0, 22) to (0.25, −2).

b) Noise insulation; less massive.

8 a) There are various ways to calculate this; e.g. $pV = NRT; N = 100000 \times 24 \times \frac{3}{8.31} \times 273 = 3.2 \times 10^3$ moles.

b) Energy required $= 3.2 \times 10^3 \times 30 \times 22 = 2.1 \times 10^6$ J ; 3.6 MJ costs 8 p, so cost is 4.7 p.

c) (i) Surface area to outside $= 30$ m^2, cost per hour $= 30 \times 41 \times 3600 \times \dfrac{8}{3.6} \times 10^6 = 9.8$ p.

(ii) Cost per hour $= 30 \times 14 \times 3600 \times \dfrac{8}{3.6} \times 10^6 = 3.4$ p.

d) It is not reasonable; once there is a temperature difference, then energy loss occurs, e.g. when the temperature is 9 °C (half way) the additional cost per hour will be 1.7 p with cavity fill.

9 There are a number of possibilities to write about: perhaps capacitor discharge and radioactive decay, or, alternatively, gravitational and electric fields.

CHAPTER NINE

MAKING MORE CONNECTIONS

Comprehension questions

The number of marks for each question is shown in parentheses at the end of the question.

Comprehension 1 – Orbiting generator
(time allowed – 40 minutes)

In 1996 NASA scientists undertook what was to prove to be an unsuccessful space mission in an attempt to generate electrical energy in space. They planned to release a satellite of mass 500 kg tethered by a metal wire from a space shuttle in close Earth orbit (200 km from the Earth's surface) and travelling at 8 km s^{-1}. Over a period of some six hours the satellite would travel a distance of 20 km away from the shuttle using a gas thruster to keep the tether taut. Eventually the tether was to be stretched between the shuttle and the satellite along a radial line directed towards the centre of the Earth.

The metal wire would thus be swept through the Earth's magnetic field and a voltage would be generated between its ends. The tether had the appearance of a white shoelace, but was in fact made from copper and Kevlar. Kevlar is a polymer that is stronger and tougher than most metals with a low electrical conductivity and high corrosion resistance. A potential difference of 5 kV was expected between satellite and shuttle, the satellite being charged positively. This would lead to free electrons in the upper atmosphere being attracted to the satellite. The circuit would be completed by an electron gun on the shuttle firing electrons back into the atmosphere again. Preliminary estimates indicated that 1 kW would be generated.

Unfortunately the Kevlar–copper tether snapped during deployment and the satellite eventually ended up spinning some 3000 km away from the shuttle. The tether was 2.5 mm thick, more than strong enough to hold the satellite which should have exerted no more than 50 N on the tether whilst in space.

A later investigation concluded that the failure was probably due to 'arcing and burning in the tether leading to a tensile failure after a significant portion of the tether had burned away'. Shards of metallic aluminium and silver, remnants of the manufacturing process, were discovered in the tether and these probably punctured the insulating Kevlar layer whilst it was stored on its reel.

Mass of Earth = 6×10^{24} kg; e, charge on electron = -1.6×10^{-19} C
$G = 6.7 \times 10^{-11}$ Nm^2 kg^{-2}; radius of earth = 6.4 Mm

1 **a)** (i) What provides the force to keep the shuttle in orbit?
    (ii) Show that the values in the passage for shuttle speed are correct.   *(3)*

   b) The tether needs to be kept along a radial line from the centre of the Earth and thrusters are used to achieve this. Explain why the difference in orbital radii between the shuttle and the satellite results in a drift of the tether.   *(2)*

   c) Calculate the difference in speed between the shuttle and the satellite when the tether length is 20 km.   *(3)*

2 **a)** Explain why both copper and Kevlar are included in the design of the tether.   *(2)*

   b) Explain why burning of the Kevlar could have lead to tensile failure of the tether.   *(1)*

3 **a)** Explain why there is a movement of electrons in the conducting tether and hence why a voltage is generated as the tether cuts the Earth's magnetic field.   *(5)*

   b) Estimate the minimum magnetic field that is required to produce the induced emf of 5 kV.   *(2)*

   c) Explain the consequences of *not* firing the electrons back into space with the electron gun.   *(2)*

   d) (i) Estimate the current required to generate 1 kW of power.
    (ii) Estimate the number of electrons that must be fired into space every second.   *(3)*

4 The mission astronauts were trained to prevent the satellite bobbing in and out as if on a spring.

   a) Discuss the variation of tension in the tether that would result from this motion.   *(2)*

   b) The last part of the process of reeling the satellite back into the shuttle loading bay is the riskiest. This is because, as the tether shortens, the tension variations are amplified. Explain why for a constant amplitude of bobbing motion, a shorter tether experiences greater tension.   *(3)*

Hint **1** **a)** (i) What supplies the centripetal force? (ii) Write down or derive the formula for orbital speed (equate centripetal force to gravitational force between Earth and satellite).

       b) The satellite and shuttle will be at slightly different radii, speed depends on orbital radius.

       c) Calculate either the difference (with calculus) or both speeds (with a calculator) and subtract.

   2 **a)** Think about the electrical and mechanical properties of both substances.

       b) What happens when the majority of the mechanical strength disappears?

   3 **a)** The electrons are moving in a magnetic field, this movement is what is meant by an electric current. What force acts on the electrons as a result?

       b) The equation is emf = rate of change of magnetic flux = (magnetic field strength) × electron speed × length of conductor.

       c) Consider the build-up of electrons at one end of the tether.

       d) Use power = VI and $Q = It$.

4 a) If spring-like, it will be simple harmonic, describe this. You could sketch a graph. What would be the consequence of the graph going below the *x*-axis?

b) The bobbing amplitude is constant, but the overall length is decreasing, what effect does this have on the *strain*?

Comprehension 2 – Weightless towers (time allowed – 45 minutes)

The first manned spaceflights of the 1960s allowed many experiments to be performed in weightless conditions. Convection and other buoyancy effects caused by density difference disappeared and additionally the influence of weak intermolecular forces such as those that cause surface tension in liquids was allowed to dominate. This proved invaluable in the study of many phenomena. The development of *drop towers* has allowed researchers to recreate similar conditions on Earth.

In the strict sense, true 'weightlessness' can never be attained in a gravitational field. However, it can be simulated inside a capsule that is allowed to fall freely. During free fall the reaction forces between the contents and the floor of the capsule are zero as they would be in the absence of a gravitational field.

In September 1990, the drop tower at Bremen was commissioned as the first weightless laboratory of this type in Europe and one of the largest in the world. Capsules fall 110 m from the top of an evacuated drop tube. They are then decelerated in a special chamber at the base of the tower. The capsules, which are airtight cylinders 2 m long, carry an experimental payload of 180 kg in addition to their on-board electronics and computer system. Data from the falling capsule are transmitted to the control centre via a modulated laser beam at a rate of up to 4 Mbits s^{-1}. At the end of the free fall the capsule enters an 8 m high deceleration tank which is filled with polystyrene granules. The kinetic energy of capsule is dissipated by the inelastic deformation of the polystyrene particles. The tank is designed to limit the maximum deceleration of the capsule to 30 *g* thus enabling normal laboratory equipment with shockproof construction to be used in the capsule.

Any forces other than gravity acting on the capsule would create disturbing differential accelerations between the capsule and its contents and so destroy the weightless conditions. Since it is almost impossible to eliminate external forces entirely, researchers prefer to speak about the creation of a microgravity environment rather than true weightlessness.

In the 'Bremen' drop tower the aim has been to reduce these unwanted accelerations to about 1×10^{-5} m s^{-2}. Air resistance on the falling capsule is the major factor. The drag force D acting on the capsule is given by the equation:

$$D = 0.4\rho v^2$$

where v is the velocity of the capsule and ρ is the density of the air. In order to achieve the above target it is necessary to use a powerful pumping system to reduce the air pressure in the tower to 1 Pa (normal atmospheric pressure is about 100 000 Pa).

Disturbing accelerations can also arise if the capsule spins during the drop. For a 2 m long capsule the rotation rate about a horizontal axis must be no more that 10^{-2} rad s^{-1} if the target acceleration is not to be exceeded inside the experiment bay. Special efforts have been made to achieve this: a free-standing shield protects the tower from horizontal wind-induced forces, which could disturb the capsule prior to release and a novel drop mechanism is used which releases the capsule rapidly and symmetrically.

Electric and magnetic fields can also affect the capsule. Care is taken to shield the capsule from the Earth's magnetic field and also to prevent the build-up of static charge on either the capsule or tower. Experimental equipment inside the capsule, such as current-carrying coils or permanent magnets may induce currents in the steel wall of the tower as the capsule falls causing consequent deceleration.

The tower has been widely used by materials scientists researching into the solidification of metals since the absence of buoyancy effects allows special metallic alloys to be produced. Similarly the elimination of convection currents suppresses early crystallisation making it possible to supercool metals. Special kinds of glass have been produced in this way. Another area of research has been into combustion. The way in which a substance burns is normally influenced greatly by thermal convection. When a fuel is ignited under microgravity conditions a spherical flame shape arises. This simplifies the numerical treatment of the problem and the influence of other factors such as the air/fuel ratio becomes more evident. Because combustion is such a fundamental process in heating systems and engines, the drop-tower experiments have contributed significantly to improving fuel economy and reducing pollution throughout the world.

1 a) What does the author consider to be true *weightlessness*? *(2)*

b) Would the author use the word *true* or *simulated* to describe the weightless conditions experienced in a space station orbiting the Earth? Explain your answer. *(2)*

2 a) Show that the weightless conditions last about 5 s in the drop tower (take g to be 9.8 m s^{-2}). *(2)*

b) Calculate the speed of the capsule just before it enters the deceleration chamber. *(2)*

c) Calculate the mean deceleration of the capsule assuming that it is slowed uniformly over the 8.0 m of the deceleration chamber. *(2)*

d) In fact the deceleration is unlikely to be uniform. Sketch a graph showing how the deceleration of the capsule might vary with time from 0.1 seconds *before* entering the deceleration chamber until it finally comes to rest. *(2)*

3 **a)** The density of air at normal atmospheric pressure is about 1.3 kg m^{-3}. Show that the maximum drag force experienced by a capsule as it falls in the evacuated tower is about 1×10^{-2} N. *(2)*

b) The total mass of the capsule and payload is 300 kg. Calculate the effective maximum gravitational field strength that would result from this drag force. *(2)*

c) The pumping system fails and allows the air pressure in the tower to return to normal atmospheric value. State and explain whether the falling capsule could reach its terminal speed under these conditions. *(4)*

4 **a)** The presence of a current-carrying coil in the falling capsule would affect the microgravity environment. Explain why. *(5)*

b) Part of one of the experiments in the capsule consists of a bar of copper 0.40 m long held horizontally. Calculate the emf that would be generated in the bar if the Earth's magnetic field were allowed to act on the contents of the tower (take the horizontal component of the Earth's field to be 70 μT). *(2)*

Hint **1** **a)** and **b)** How can something have no weight at all? Why is orbital motion around the Earth not truly weightless?

2 **a)** Use $s = ut + \frac{1}{2}at^2$; **b)** use $v = u + at$; **c)** use $v^2 = u^2 = 2as$; **d)** most of the deceleration will be at the end of the time as the polystyrene begins to crush. You should give approximate values for the maximum deceleration and time taken for the capsule to stop, remember to label axes fully.

3 **a)** Use equation in the passage; **b)** $g = F/m$; **c)** terminal speed occurs when drag force = weight; compute speed for this to occur.

4 **a)** Coil current 'carries' a magnetic field with it. This is moving relative to the wall of the tower. What happens if the tower wall has a conductor in it?

b) emf = Blv.

Comprehension 3 – Satellites for TV (time allowed – 60 minutes)

The sight of a satellite dish on a house is now a commonplace, but the physics of the satellite and the nature of its communication with the receiving station on the Earth are still intriguing. In the early days of television relay, the communication links were handled at a national level so that a satellite such as *Telstar* (which was one of the first to relay television signals between Europe and North America) radiated signals to the receiving station at Goonhilly Down in Cornwall. Communication times for each receiving session were brief with only a short time between the acquisition of signal and the disappearance of the satellite over the horizon and the accompanying loss of signal. The signal was then amplified and relayed by landline to national television studios where it was mixed into the programme before transmission to the transmitters.

Nowadays such an arrangement seems crude to us with the satellites positioned permanently above our heads and with a 24 hour per day availability of television signals.

The early *Telstar* was in a close-Earth orbit, about 160 km above us. The modern satellites are positioned about 35 000 km from the Earth in a geostationary orbit. This has two disadvantages for the satellite operator: more energy is required to place each kilogram of satellite in orbit at the geostationary radius and the electromagnetic power reaching the Earth from the antenna on the satellite is weakened much more, given the greater distance the signal has to travel.

This means that great attention is paid to the antenna dish used to transmit the signal. The requirement is for the signal to be radiated to a region roughly 1000 km wide. A high frequency of transmission is used (12 GHz) so that the size of the dish can be reduced. The radiated powers of many satellites are of the order of 100–1000 W and this needs special arrangements at the receiver in order to produce sufficient input signal power to operate correctly. Other overheads in the satellite mean that the total power consumption of a satellite can easily approach 2 kW.

An output power of this size needs considerable energy consumption by the satellite. There are a number of possible power systems, but a common one is the solar cell. A possible design for a satellite is to have two banks of solar cells with the dish-shaped aerial between them transmitting a signal of wavelength 0.1 m. The cells have a conversion efficiency of about 15%. But even with the radiation power from the Sun at 1400 W m^{-2} the cells have such a low efficiency that cells with substantial mass and bulk are needed to supply the needs of the satellite.

Mass of Earth = 6.0×10^{24} kg; universal gravitational constant $G = 6.7 \times 10^{-11}$ N m^2 kg^{-2}; radius of Earth = 6.4 Mm; speed of light $c = 3.0 \times 10^8$ m s^{-1}; Avogadro's number = 6.0×10^{23}.

1 This question is about the orbital characteristics of close-Earth orbit and geostationary satellites.

   a) Explain what is meant by a geostationary orbit and state the orbital period of the satellite. *(2)*

   b) (i) Show that the orbital radius of the geostationary satellite is about 42 000 km.
      (ii) Explain the discrepancy between this result and the distance quoted in the passage. *(3)*

   c) A geostationary satellite has a mass of 180 kg. Calculate the energy needed to place it in geostationary orbit about the Earth. *(2)*

   d) There is a proposal to place a geostationary satellite over Bristol. Explain with the aid of a diagram why this is impossible. *(3)*

   e) Explain why the 'communication times for each receiving session were brief' for *Telstar*. *(2)*

2 This question is about the radiating properties of the satellite antenna.

   a) Calculate the wavelength of the electromagnetic wave. *(2)*

   b) State the region of the electromagnetic spectrum to which this radiation belongs. *(1)*

   c) Calculate the diameter of the satellite dish that is required to illuminate a region of the Earth's surface 1000 km across at a distance of 35 000 km. *(3)*

   d) The satellite transmits a power of 1.0 kW. Calculate the power received by a satellite dish receiver of area 0.25 m² at the Earth's surface. *(2)*

3 This question is about the power required by the satellite and its generation by solar cells.

   a) What area of solar cells is required in order to generate the quantity of power required for the satellite to function correctly? *(2)*

   b) What are the drawbacks of solar power for the satellite? *(2)*

4 This question is about an alternative power source for the satellite that derives energy from the decay of a radioactive isotope.

   a) A possible power source is radium-228 which has a half-life of 1600 years. Show that the decay constant for this isotope is about 1.4×10^{-11} s^{-1}. *(2)*

   b) The radium decays to form actinium-228 (Ac). Complete the equation for this decay.

$$^{226}_{89}\text{Ra} \rightarrow$$ *(3)*

   c) The actinium goes on to decay further. Each radium atom produces, on average, 5 particles, each of energy 10^{-12} J. 50% of this alpha-particle energy emitted in the decays can be converted into electrical energy. Calculate the mass of radium that is required to power the satellite. *(4)*

Hint **1 a)** Think about the satellite and its speed relative to the Earth if it is to be stationary.

   b) **(i)** You know the orbital period of the satellite, equate Newton's law of gravitation and the centripetal force on the satellite to evaluate *r*. The time period is equal to the circumference of the orbit divided by the orbital speed.
   (ii) What is the difference between the values and what could this be due to?

   c) Use the formula for gravitational potential energy (gpe) $\left(\frac{GMm}{r}\right)$ and calculate the difference between gpe at the surface at gpe at 42 000 km.

   d) What does the centripetal force direction need to be in order to orbit over Bristol; what does the gravitational force provide?

   e) *Telstar* was much closer to the Earth, what can you say about its orbital time therefore?

2 a) Use $c = f\lambda$.

   b) Look this up if you need to – you may be required to know these by heart.

c) This involves some diffraction theory; the diffracting angle out to the first minimum is given by $d\sin\theta = n\lambda$, where $n = 1$. The order to carry this out is (i) calculate θ knowing the 'footprint' required and the distance to the surface (remember that θ is from the centre of the pattern to the first minima); (ii) use this value of θ to calculate d which is the diameter of the dish.

d) The energy is spread across a circle 500 km in radius. Work out the power per square metre at a distance of 36 000 km. How much of this energy is fed into 0.25 m²?

3 a) Use data in the passage to estimate the area. Do not forget the efficiencies quoted in the passage.

b) Think about what could make the satellite be in the dark.

4 a) Use the relationship between half-life and decay constant (decay probability): $t_{\frac{1}{2}} = \ln 2/\lambda$.

b) The alpha particle has 4 nucleons including 2 protons. The numbers at the top and bottom of the equation should balance.

c) Find how many alpha particles need to decay each second to produce the correct amount of energy. Each radium atom yields five particles, so you know dN/dt in the decay equation $dN/dt = -\lambda N$ to work out N, then Avogadro's number will give you the number of moles and hence the mass knowing that the nucleon number in grams is one mole.

Comprehension 4 – Inside a star (time allowed – 60 minutes)

The balance between gravitational attraction and thermal pressure plays the principal role in determining the structure of a star. The study of stellar structure began with much discussion about the physical state of matter in stars. It was thought that stars could not be solid because their temperatures were so high. Equally, stars could not be gaseous because their mean densities were too high (e.g. at a typical point in the Sun the temperature is about 2×10^6 K and the concentration of particles is about 2×10^{30} m^{-3}). It is now believed that stars are composed of an almost-perfect gas in most circumstances. This almost-perfect gas is unusual in two ways.

The more important respect is that the stellar material is an ionised gas or plasma. The temperature inside stars is so high that all but the most tightly bound electrons are separated from the atoms. This makes possible a very much greater compression of stellar material without deviation from the perfect gas law because a nuclear dimension is 10^{-15} m, compared with a typical atomic size of 10^{-10} m. The word *plasma* is a name given to a quantity of ionised gas. It has been recognised in recent years that a plasma can be regarded as a fourth state of matter and that most of the material in the universe is in this fourth state. It differs from an ordinary gas because the forces between electrons and ions are considered to have a much longer range than the forces between neutral atoms.

The second important difference between most laboratory conditions and the interiors of stars is that radiation is in thermal equilibrium with matter in stellar interiors and its intensity is governed by Planck's law. Particles in a gas exert a pressure that can be calculated from the kinetic theory of gases by considering collisions of particles with an imaginary surface in the gas. Similarly, photons exert a pressure known as radiation pressure. If β is the fraction of the total pressure P contributed by gas pressure towards opposing gravitational contraction then $(1-\beta)$ is the fraction contributed by radiation pressure. It was once thought that radiation pressure was of comparable importance to gas pressure in ordinary stars. It is now realised that although there are some exceptional stars in which radiation pressure is of vital importance, it is only of marginal importance in most stars.

From the kinetic theory of gases, the pressure of a perfect gas can be shown to have the form:

$$P_{gas} = nkT$$

where n is the number of particles per cubic metre and k is Boltzmann's constant $(1.4 \times 10^{-23} \text{ J K}^{-1})$. This expression for the pressure can be made to correspond with the usual form for Boyle's law as follows. If we consider a mass of gas m of molecular weight M, which occupies a volume V, its pressure P_{gas} is given by:

$$P_{gas} V = (m/M) RT = (m/M)N_A kT$$

where R is the gas constant $(8.3 \text{ J mol}^{-1} \text{ K}^{-1})$, N_A is the Avogadro constant $(6.0 \times 10^{23} \text{ mol}^{-1})$ and $k = R/N_A$.

If we consider a cubic metre of gas and note that $n\ (= mN_A/M)$ is the number of particles in a cubic metre, the equation $P_{gas} = nkT$ results. The corresponding expression for radiation pressure is:

$$P_{rad} = \frac{1}{3} a\, T^4$$

where a is the radiation density constant $(7.55 \times 10^{-16} \text{ J m}^{-1} \text{ K}^{-4})$.

So what is the origin of the energy that maintains the material in the star in the plasma state? Stars generate energy for most of their lives by combining protons in a process as a result of which helium nuclei are formed. The energy release keeps the star hot enough for fusion to operate, but ultimately it escapes from the surface, and this is what causes the star to shine. Knowledge of the power radiated by a star can lead to a determination of the power it is generating if the size of the star and its surface temperature are known. The controlling relationship resembles the radiation pressure equation and is:

$$W = \sigma A T^4$$

where W is power radiated from the surface of area A and temperature T. σ is the Stefan-Boltzmann constant (6×10^{-8} W m^{-2} K^{-4}).

Additional data:

Radius of Sun	7.0×10^8 m
Mass of Sun	2.0×10^{30} kg
Temperature of visible surface of Sun	6.0×10^3 K
e charge on electron	-1.6×10^{-19} C
One year	30 Ms
$1/4\pi\varepsilon_0$	9.0×10^9 N m^2 C^{-2}
Mass of a proton	1.7×10^{-27} kg

1 State the names of:

  a) the force that is compressing stellar material and

  b) the force that prevents the complete collapse of the star. *(2)*

2 Explain what is meant by the phrase 'stars are composed of an almost perfect gas'. *(2)*

3 Explain how plasma – the fourth state of matter – is produced and how it differs from the other three. *(2)*

4 What is meant by the *range* of a force. Illustrate your answer with a sketch graph of the two cases of **a)** electrostatic forces between an electron and a proton, and **b)** gravitational forces between an electron and a proton. *(3)*

5 Estimate the root mean square (rms) speed of protons in the Sun. *(3)*

6 Compare numerically the radiation pressure and gas pressure at a typical point in the Sun. Comment briefly on your answer. *(3)*

7 It can be deduced from the passage that $P = \frac{\sigma T^4}{3(1-\beta)}$. Using similar reasoning, deduce another equation for P and combine these to obtain a total pressure relationship in which temperature does not appear. *(2)*

8 Show that the power *radiated* by the Sun is about 5×10^{26} W. *(2)*

9 Calculate the power generated by one kilogram of the Sun. *(2)*

10 When four protons combine to form one helium nucleus, 4×10^{-12} J are liberated. How many protons are combining each second in one kilogram of the Sun? *(3)*

11 Assuming that the Sun presently consists of 75% hydrogen and 25% helium by mass, calculate the number of hydrogen nuclei per kilogram in the Sun. *(2)*

12 Assume that the power generation has been constant throughout the Sun's life and that it contained 100% hydrogen originally. Estimate the age of the Sun. *(3)*

13 Two protons cannot fuse unless they overcome their electrostatic repulsion and approach closer than 1.5×10^{-15} m.

a) Calculate the energy required to bring the protons together. *(2)*

b) The average thermal energy of an atom is approximately equal to *kT*, where *k* is the Boltzmann constant. Assume that this energy alone is sufficient to cause two atoms to fuse. Show that the assumption that fusing protons are brought close enough by thermal energy alone leads to a typical interior Sun temperature of about 10^9 K. *(2)*

c) The result in (b) is very different from the temperature estimate quoted in the passage. Suggest why fusion occurs at the lower temperature. *(2)*

14 Discuss, in as much detail as possible, the effect on the Sun's behaviour if the electrostatic energy needed to bring two protons together were suddenly reduced by a factor of 10. *(5)*

Comprehension 5 – Black holes
(time allowed – 40 minutes)

Black holes continue to hold the attention of both the general public and the astronomical world. For the layman there are the frequent references to worm holes, black holes and so on in science-fiction, for the astronomer there is a continuing need to understand the complex mathematics of the black hole.

So what is the origin of a black hole? As stars age, they use up energy resources and, as a result, can shrink to diameters very much smaller than their size throughout most of their lives. Our own Sun has a radius of 7×10^8 m with a density of 1.4×10^3 kg m^{-3}, but if the Sun could shrink until its density is similar to that of an atomic nucleus, then it would have a radius of about 10 km. Such a dense star is called a *neutron* star. In fact, the Sun will not progress beyond this point to become a black hole because its mass is too small. A black hole has an escape velocity that is equal to the speed of light so that nothing, even light itself, can escape from the hole's surface. Black holes are hard to observe and can, in principle, only be seen by their effect on nearby matter. A black hole attracts particles that attain very high speeds as they fall into the hole itself.

Neutron stars show a similar phenomenon. The gravitational field of the dense star is so great that it can 'tear off' the outer layers of a nearby larger star (often one in a binary system with the neutron star). These particles are accelerated to enormous speeds and, as they approach the surface of the neutron star, become intensely hot and therefore radiate electromagnetic energy that can be detected by astronomers.

---

1 a) Explain what is meant by the term *escape speed*.

b) Show that the escape speed v_e of an object on the surface of a spherical body of mass M and radius R is given by:

$$v_e = \left(\frac{2GM}{R}\right)^{1/2}$$

where G ($= 6.7 \times 10^{11}$ Nm^2 kg^{-2}) is the gravitational constant. *(4)*

2 a) Show that the mass of the Sun is about 2×10^{30} kg.

   b) Calculate the escape velocity from the visible surface of the Sun. *(4)*

3 Calculate the radius that the Sun would need to have in order for it to become a black hole (speed of light, $c = 3 \times 10^8$ ms^{-1}). *(2)*

4 The passage describes how matter falling towards a neutron star can attain high speeds and hence radiate energy.

   Show that the kinetic energy attained by one mole (2.0×10^{-3} kg) of hydrogen gas falling onto the surface of the Sun, if the Sun had a radius of 12 km, is about 2×10^{13} J. *(3)*

5 Estimate the temperature reached by the gas if all of this kinetic energy is used to heat the gas.

Molar gas constant = 8.3 J mol^{-1} K^{-1} *(2)*

6 A rough estimate of the wavelength of most energetic photons emitted by a hot gas at a temperature T can be found by equating the energy of the photon to kT, where k is the Boltzmann constant (1.38×10^{-23} J K^{-1}).

   a) Estimate the smallest wavelength of a photon that is emitted by a gas of temperature 10^{12} K (Plank's constant, $h = 6.6 \times 10^{-34}$ Js).

   b) State the region of the electromagnetic spectrum in which this radiation lies. *(3)*

7 It is possible for two gamma-ray photons of equal energy to interact and for their energy to be transformed into an electron (mass 9.1×10^{-31} kg) and its antiparticle the positron. A positron has the same mass as an electron with an opposite charge. Calculate the smallest energy in electron-volts that each photon must have for this interaction to take place. *(3)*

Comprehension 6 – Mobile telephone technology (time allowed – 45 minutes)

The past few years have seen an explosion in the use of mobile telephones. Radio is a convenient and accessible method for mobile communication. Cables are not used for connection between stations, the waves are invisible and inaudible to other people, and the user does not need to be within sight of a transmitter. However, it has some problems of implementation, one being how to manage the large number of simultaneous signals being transmitted over the system.

The available spectrum

Mobile phones do not transmit directly between handsets. Each mobile is in contact with the local base station, each with its own antenna (aerial). Connections are created from mobile to base, base to central exchange, and then from the exchange via a second base station to the receiving mobile. The present provision is for two frequency bands, one from 890 MHz to 915 MHz and another from 935 MHz to

960 MHz. The lower band is used for transmission from the mobile telephones to the base stations, the higher band for transmission from base to mobile.

So, each mobile telephone operates on two frequencies: one for transmission somewhere between 890 and 915 MHz, the other for reception somewhere between 935 and 960 Mhz. The transmitter and receiver do not however operate on one unique frequency. The speech information occupies a band of frequencies roughly 3–4 kHz wide (this is known as the bandwidth) and there are other control signals to be transmitted too. Each complete signal is about 20 kHz wide and the individual channels are set 25 Hz apart to avoid overlap. Hence, there are only a relatively small number of available channels to share out amongst the hundreds of millions of mobile telephone users. How is this achieved?

Path loss, normally a major disadvantage in radio communications, is the phenomenon used to allow apparently limitless numbers of users to use the network without overload. The most important factor in path loss is the inverse square law. Power density D is defined as the power falling on a receiving aerial covering an area of 1 m^2 perpendicular to the incident radiation.

Loss caused by the inverse-square law can sometimes be compensated by making the transmitting aerial directional, so that it transmits more power in some directions than in others. Thus less power is wasted in unwanted directions. However, in mobile radio it is difficult by definition to make the transmitting aerials directional.

Frequency re-use

Path loss can be a serious problem in some communication systems because the quality of service degrades with decreasing signal strength, but in mobile phone communication it is an advantage because it allows frequency re-use. This is possible because the signal from one user will be greatly reduced in intensity by the time it reaches the location of the other user, and will therefore cause negligible interference.

Imagine two transmitters, A and B, operating on the same frequency at the same signal power. A mobile unit is tuned to transmitter A and its distance to B is ten times bigger than the distance to A. This means that the signal strength at the unit from A will be 100 times larger than the signal from B and there is no interference from the B unit. The B unit can be used by a mobile user closer to B on the same frequency without any co-channel interference.

The minimum permissible distance between the outer edges of transmitting zones using the same frequency is called the re-use distance. It depends on the maximum distance of the mobile from the wanted signal transmitter and on the strength of the interfering signal that can be tolerated. Suppose that the re-use distance is 50 km. The service area can be imagined to be divided into circles of radius 25 km with a transmitter at the centre of each. The transmitters would then be separated by 50 km, which is the required re-use distance.

The service area is divided into a network of cells each served by one base station. The same frequencies may be assigned to different cells, but only if they are sufficiently far apart for frequency re-use to be permissible. There is always at least the width of one cell between two cells tuned to the same frequency, this guarantees a minimum re-use distance.

1 State the merits of radio waves as the communication medium for mobile phones. *(3)*

2 Explain, *with reference to the physical principles involved*, why it is not necessary for the mobile phone user to be able to see the transmitter? *(2)*

3 Calculate the mean transmission wavelength assigned to mobile phones for transmission to a base station in Europe. *c* the speed of electromagnetic waves in a vacuum is 3×10^8 m s^{-1}. *(2)*

4 Estimate the number of mobile users possible on the system in each direction with a channel spacing of 25 kHz. *(2)*

5 **a)** State the effect that makes the greatest contribution to path loss.

b) Write down a relationship that connects *D* the power density, *P* the power of the transmitter in watts and *r* the distance from the transmitter to the aerial.

c) State another factor that will contribute to path loss and explain how it can arise.

d) Calculate the signal power received by an aerial of effective area 30 cm^2 when placed 25 km from a transmitter of power 3.0 kW. *(6)*

6 **a)** Explain what is meant by the re-use distance.

b) State the factors on which re-use distance depends.

c) In Figure 9.1, the straight line AB passes through three of a long line of adjacent equal circular cells of diameter *x*, with transmitters T of equal power at the centre of each. Sketch a graph to show how the power density varies with the distance from A along the line AB.

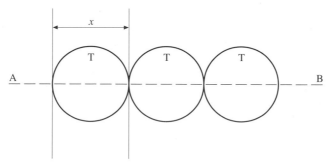

Figure 9.1

d) A re-use distance of 50 km is adopted in a cell network using identical transmitters. Show that the power density of the signal from the nearest interfering transmitter will be about 10% of the power density from the wanted transmitters. *(7)*

Data analysis questions

Data analysis 1 – Windmills in the sea (time allowed – 60 minutes)

These questions are about the use of collections of wind turbines built out to sea that supply energy for domestic use in the UK.

Data

Density of air	1 kg m^{-3}
Density of copper	$9.0 \times 10^3 \text{ kg m}^{-3}$
Density of aluminium	$2.7 \times 10^3 \text{ kg m}^{-3}$
Resistivity of copper	$1.7 \times 10^{-8} \ \Omega \text{ m}$
Resistivity of steel	$10 \times 10^{-8} \ \Omega \text{ m}$
Resistivity of aluminium	$2.9 \times 10^{-8} \ \Omega \text{ m}$
Area of sea 30 km from coast at suitable depth	4000 km^2
Capital cost of one wind turbine generating 2.5 MW	£600 per kW; construction of tower represents 40% cost
Wind turbine designed for 1.5 v_{m} will operate for about 40% of the time	
Cost of cables of length 30 km	£50 per kW
Number of seconds in one year	3×10^7
Magnetic field strength around a long straight wire	$\mu_0 I / 2\pi r$
μ_0 permeability of free space	$4\pi \times 10^{-7} \text{ H m}^{-1}$

1 Wind with an air density of ρ and speed v blows towards a wind turbine of area A.

    a) Show that the mass of air delivered to the turbine every second is $\rho A v$.

    b) Hence, show that the wind delivers a power of ½ $\rho A v^3$ to the turbine. *(3)*

2 Wind speed is very variable and engineers take the mean power generated by the turbine to be $\frac{1}{2}k\rho A v_{\text{m}}^3$, where k is a constant and v_{m} is the mean speed throughout the year. Figure 9.5 shows wind–velocity data for one year at a particular site.

    Show that the value of k is roughly 2.5. *(4)*

3 Wind speed varies with height. For wind over a flat terrain, the wind speed v_H at height H varies according to:

$$V_H = CH^{0.15}$$

where C is a constant.

    a) Use this information to show that there is an advantage in building a high structure for a wind turbine.

    b) Suggest another factor that you would need to take into account in order to decide on the best building height for a wind turbine.   *(4)*

4 Estimate the total blade area for one wind turbine that is to have the following characteristics:

    ★ v_m for the region is 9 m s^{-1} at 50 m height

    ★ wind turbine to be 50 m high

    ★ operational speed $1.5v_m$

    ★ 2.5 MW to be generated.   *(3)*

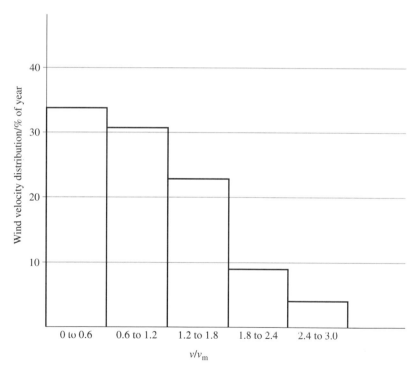

Figure 9.2 Wind velocity distribution

5 Comment on the extent to which the blade area that you estimated in Question 4 is practicable. *(3)*

6 Wind turbines in clusters (a wind farm) cannot be placed too close to each other. The minimum separation must be seven times the largest blade dimension. Explain why. *(2)*

7 Estimate the cost of electrical energy generated by a cluster of 50 wind turbines designed to produce a mean power of 125 MW. The capital cost is to be met by a loan on which interest has to be paid at 5% per year.   *(2)*

8 The electricity is to be carried to the land through two copper cables each of length 30 km and copper diameter 0.2 m.

   a) Calculate the resistance of these cables.

   b) The electrical energy is transmitted at a potential difference of 432 000 V. Show that there is an electric current of about 1 kA in the cables.

   c) Calculate the loss of electrical energy in the cables.   *(6)*

9 Calculate the weight of copper in each cable.   *(2)*

10 The cables are redesigned and the copper is replaced by an aluminium conductor surrounded by a steel sheath (see Figure 9.3).

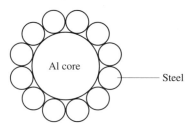

Figure 9.3 Cross-section of aluminium – steel cable

   a) Explain why the copper has been replaced by aluminium.

   b) Explain why the aluminium has been sheathed with steel.

   c) Calculate the electrical resistance of the combination cable. The diameter of the aluminium core is 0.2 m and each of the 12 steel cables is of diameter 0.07 m.   *(5)*

11 The cables are 1.5 m apart.

   a) Calculate the magnetic force of attraction between them per metre length.

   b) State whether this force is attractive or repulsive.   *(5)*

Hint   **1**   a) Imagine a horizontal column of air equivalent to one second of travel, with an area the same as the turbine, hitting the turbine, what mass of air is in this column?

   b) apply the same idea to the kinetic energy.

2   For each bar on the graph determine the fractional contribution to $\frac{1}{2}\rho Av^3$ at the mean speed of the bar range (i.e. there is a contribution of $0.33 \times \frac{1}{2}\rho A \times 0.3^3$ from the first bar).

3   a) Evaluate the power output using the equation in the previous question.

   b) What else increases as the height goes up.

4   Use data and relationships already determined to compute the required area.

5   What would the dimensions of a square with this area have to be?

6 How much moving air would be received by a turbine immediately behind another?

7 Use data from question to compute cost.

8 **a)** $R = \rho l/A$ **b)** power $= VI$ (c) I^2R.

9 weight $=$ mass $\times g$; mass $=$ volume $\times$ density.

10 **a)** and **b)** compare density and mechanical properties of Al and Fe; **c)** all cables are in parallel; compute resistance of Al, resistance of one strand of Fe, hence total resistance.

11 Compute magnetic field strength of one cable at the centre of the other; then compute magnetic force on this second cable.

Data analysis 2 – Power boat performance (time allowed – 45 minutes)

These data analysis questions are about some of the factors that affect the performance of a small boat driven by an engine. Some data about the boat and its power unit are provided below.

Total mass of boat including its single boatman: 500 kg

Table 9.1 shows how the power developed by the engine varies with the speed of the engine.

Engine speed/ revolutions per minute	500	1000	1500	2000	2200	2400	2600	2800	3000	3500	4000
Output power of engine/kW	1.2	2.3	3.4	4.4	4.5	4.5	4.4	4.2	4.0	3.5	3.0

Table 9.1 Power developed by the engine

Table 9.2 shows how the drag force acting on the boat varies with total mass of the boat (boat + occupant + cargo).

Total mass of boat + cargo/kg	Drag force/N at a boat speed of		
	1 m s^{-1}	2 m s^{-1}	3 m s^{-1}
500	200	290	380
800	240	330	460
1100	290	400	550
1400	350	480	660
1700	430	600	830

Table 9.2 Drag force acting on the boat at various speeds

> There are two propeller systems:
>
> ★ system A rotates at the same speed as the engine;
>
> ★ system B rotates at twice the speed of the engine and has a different design.
>
> The thermal efficiency of the engine is 20%.
>
> | | |
> |---|---|
> | Fuel tank capacity | 10 litres |
> | Density of fuel | 0.8 kg l^{-1} |
> | Energy released when fuel is burnt | 40 MJ l^{-1} |
> | Cost of fuel | 80 p l^{-1} |

1 **a)** Use the data to construct a table showing the values of the theoretical power p required at the propeller to drive boats of different masses at different speeds.

   b) Display your data on a single graph of p against total mass (with a common set of axes).   *(6)*

2 Use the graph to estimate:

   a) the power output p required to drive the boat at 2 m s^{-1} when it carries a load of 500 kg;

   b) the largest load that can be carried at a speed of 3 m s^{-1} with power output p of 2.2 kW;

   c) the speed that can be attained with an engine power output of 1.5 kW and a load of 1000 kg.   *(3)*

3 Draw a graph of output power of the engine (y-axis) against engine speed (x-axis).   *(3)*

4 When propeller system A is used with a load of four people, 75 kg each, it travels at 2 m s^{-1} with a propeller speed of 2000 rpm. When propeller system B is used, the boat reaches the same speed with the same load but with a propeller speed of 3000 rpm.

   Compare the two propeller systems in terms of:

   a) the ratio of engine power output to theoretical power P in each case;

   b) the overall efficiency of conversion of fuel energy to useful work done;

   c) the range on a full tank of fuel;

   d) the cost of fuel per kilometre travelled.   *(9)*

Hint **1** Power = force × speed; Table 9.2 gives values of drag force for different speeds, multiplying the drag force by the speed gives the theoretical power required at that speed. Then use these derived values to plot the new graphs on the same set of axes.

   2 A series of straightforward read-offs from your new graph.

   3 Use the values from Table 9.2 to plot this graph. Draw a smooth curve through the data points.

4   Repeat the earlier questions but with the changed values. Do not forget that the two propeller systems are differently geared.

Data analysis 3 – Ozone in the atmosphere (time allowed – 45 minutes)

The atmosphere is a mixture of gases that extends to about 100 km above the surface of the Earth.

Data

Molar gas constant	$8.31 \text{ J mol}^{-1} \text{ K}^{-1}$
Avogadro constant	$6.02 \times 10^{23} \text{ mol}^{-1}$
Universal constant of gravitation	$6.67 \times 10^{-11} \text{ N m}^2 \text{ kg}^{-2}$
Radius of Earth	$6.38 \times 10^6 \text{ m}$
Mass of Earth	$5.98 \times 10^{24} \text{ kg}$
Area of Antarctic land mass	10^{13} m^2
Mean density of ozone at atmospheric pressure	2.1 kg m^{-3}

1   Table 9.3 shows the relationship between height above sea level h and the percentage of the atmosphere P that is above h.

h/km	P
0	100
20	10
40	1
60	0.1
80	0.01

Table 9.3

a)   Plot a suitable graph to display the variation of P with h.

b)   Show that P varies with h exponentially.

c)   The P against h relationship can be represented by

$$P = P_0 e^{-ah}$$

where P_0 is the percentage of the atmosphere above sea level and a is a constant.

Show that a is about 10^{-4} m^{-1}.

d)   The summit of Mount Everest is about 9 km above sea level. Find the percentage of the atmosphere that is below a climber standing on the summit.   (8)

2   The temperature 100 km above sea level (the 'top' of the atmosphere) is about 1800 K during the day.

a) Show that the rms (root mean square) speed of the oxygen atoms in this region is about 1700 m s^{-1}.

b) Show that the escape speed for the Earth is about 11 km s^{-1}.

c) A small number of oxygen atoms escape from the Earth's atmosphere. Explain how this is possible given the rms speed of the oxygen atoms and the Earth's escape speed. *(6)*

3 Ultraviolet radiation from the Sun falls on the upper atmosphere. This radiation is absorbed by the formation of ozone (O_3 molecules) and this absorption is crucial to life on Earth. Most of the atmospheric ozone is between 10 and 50 km above sea level. Ozone levels in the atmosphere above the Antarctic have been monitored for many years, some of these data are shown in Table 9.4.

Month	Monthly ozone totals/Dobson units								
	Aug	Sep	Oct	Nov	Dec	Jan	Feb	Mar	Apr
Average monthly totals 1970–80	270	270	280	340	360	350	320	300	290
Totals in 1987–88	250	180	140	180	230	320	290	260	290

1 Dobson unit is equivalent to the amount of ozone contained in a layer of ozone gas 10^{-5} m thick and area 1 m^2 at atmospheric pressure. Thus, a monthly total of 300 Dobson units means that the total amount of ozone vertically above an area of 1 m^2 is equivalent to a layer of pure ozone 3 mm thick with a temperature 273 K and at atmospheric pressure.

Table 9.4

a) (i) Plot a graph showing how the monthly values for the amount of ozone over the Antarctic in 1987/88 varied from the average values measured during the previous decade.
   (ii) Comment on the shape of this graph and explain why it gave cause for concern at the time the research was published.

b) The ozone layer becomes depleted during the time period from September to December each year. Show that the mass of ozone equivalent to one Dobson unit (DU) is about 20 mg.

c) Estimate the total mass of ozone lost during this time period in 1987/88. State **one** assumption you made in making this estimate.

d) There is a suggestion that a fleet of supersonic aircraft might transport ozone up into the Antarctic atmosphere. One aircraft could carry 10^4 kg of ozone. Comment *supporting your conclusion with a calculation* on the feasibility of this plan. *(10)*

Hint   1   a) Pressure reduces by 10 times for each 20 m increase in height. This is exponential behaviour, what graph will yield a straight line?

b) How do you test for exponential change?

c) Take logs of both sides of the equation and substitute known values into the new equation.

d) Use the equation, remember that the question asks for amount of atmosphere below the climber.

2 a) $pV = nRT = \rho$ (mean square speed).

b) Escape speed $= (2GM/r)^{\frac{1}{2}}$; substitute.

c) Do all the gas molecules have the same speed as each other?

3 a) Compute differences for each month and plot them.

b) 1 DU is amount of ozone in layer of ozone 10^{-5} m thick and 1 m² in area; density of ozone = 21 mg.

c) When is the maximum change in totals? Compute this loss over the Antarctic continent.

d) How many flights per year/per day are required? Is this number feasible?

Data analysis 4 – Froude numbers
(time allowed – 45 minutes)

Here is some data about the 'average' human:

Mass	60 kg
Height	1.7 m
Leg length	0.8 m
Minimum area of bone supporting weight in each human leg	5.0×10^{-4} m²
Breaking stress of bone	1.5×10^7 N m^{-2}
Peak total force on human leg during sprinting	$3.5 \times$ body weight
Walking speed	1.5 m s^{-1}
Sprinting speed	9.0 m s^{-1}

The gravitational field strength g is 9.8 N kg^{-1}.

$$\text{Stress} = \frac{\text{force}}{\text{area}}$$

Animals of different sizes that move in the same way as each other are said to be dynamically similar and to have the same *Froude number*.

The *Froude number* $= \dfrac{v^2}{gl}$, where v is the speed of movement of animal, l is the leg length.

The stride length s is the distance between successive prints of the same foot.

The relative stride length $= \dfrac{s}{l}$.

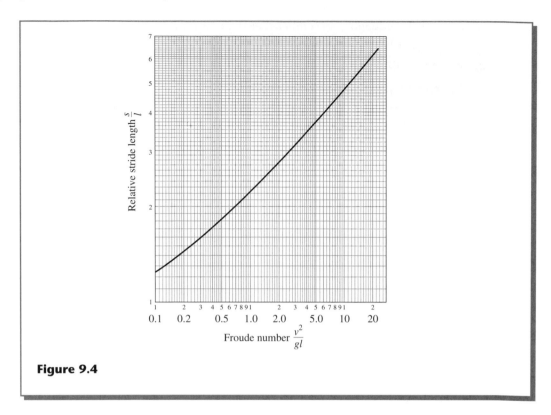

Figure 9.4

1   This question asks you to calculate the maximum size to which the human body could be increased. You should assume that the enlarged body continues to be made of the same materials with the same strength as those of the present day.

a)   Estimate the maximum stress in the bone of each leg when an average human is standing still.

b)   Suppose that each linear dimension of a human is doubled. Calculate the factor by which each of the following changes:

(i)   the area of cross-section of the bone in each leg;
(ii)   the volume of the body;
(iii) the weight of the body.

c)   Show that the maximum stress in the bone of each leg when the human is standing still is doubled when each dimension is doubled.

d)   Running imposes greater stresses on the body.

(i)   When the foot hits the ground during a sprint it is slowed, it then speeds up again. Explain how this increases the force on the leg compared to standing still.
(ii)   Use the data to calculate the maximum stress on the bone of a leg when sprinting.
(iii) Show that the greatest possible height of an enlarged human is about 6.0 m.   *(11)*

2 **a)** Calculate the Froude number for an average human:

   (i)   when walking
   (ii) when sprinting.

b) (i)   Estimate the leg length of an enlarged human.
   (ii) Hence, estimate the sprinting speed of the enlarged human. Assume that the enlarged and present-day humans are dynamically similar.

c) (i)   Explain how the graph indicates that stride length increases when a human moves faster.
   (ii) Use the graph and your answers to Question 2(a) to show that a present-day human has a sprinting stride length three times greater than the walking stride length.

d) Calculate the sprinting stride length you expect for the enlarged human.   *(12)*

3 **a)** Use the data and your earlier answers to estimate the kinetic energy (ke) when sprinting of:

   (i)   a present-day human
   (ii) an enlarged human.

b) State and explain a factor that might limit the maximum kinetic energy of the enlarged human.   *(4)*

Data analysis 5 – Towing icebergs (time allowed – 45 minutes)

Hot desert countries cannot rely on rainfall or rivers for a water supply.

One solution is to build desalination plants that remove the salt from sea water. This is an expensive solution both for the capital outlay and for the running costs.

An alternative solution is to obtain the water from Antarctic ice. Icebergs might be cut from the Antarctic ice shelves and then towed to the shores of the desert country where they would be melted and distributed.

This question asks you to compare the various options for providing water.

Data

Fresh water supply needed	2×10^6 m^3 per day
Distance from Antarctica	10^4 km
Tug towing force	5×10^6 N
Towing speed	0.8 m s^{-1} for iceberg of volume 10^8 m^3
Total cost to tow iceberg of 10^8 m^3	£18 million

Total cost to tow iceberg of 10^9 m^3	£35 million
Total running cost of desalination	£1.5 per m^3 of fresh water produced
Fuel cost of desalination	70% of total running costs
Cost of electricity	8 p per kW-hour
Energy required to melt ice at 0°C	3.4×10^5 J kg^{-1}
Sun's radiation at Earth's surface	600 W m^{-2}
Thickness of icebergs	250 m
Number of seconds in a year	3×10^7 s

1 Estimate the volume of ice required per year. *(2)*

2 Estimate the number of icebergs of each size (10^8 or m^3 or 10^9m^3) that need to be delivered to the desert country each year. *(2)*

3 Comment on which iceberg size you would advise for towing. *(2)*

4 For a 10^8 m^3 iceberg, estimate the fraction of the ice originally leaving the Antarctic that would eventually reach the desert country. *(5)*

5 Compare *quantitatively* the energy required to melt the 10^8 m^3 iceberg with the energy required to tow it 10^4 km. *(3)*

6 Discuss, with suitable calculations, the choice between allowing the ice to melt in the sun and melting it using electrical energy. Assume that a 250 MW generating station is available nearby. *(4)*

7 Compare the daily costs of desalination with those of electrical melting (with no energy input from the Sun). *(2)*

8 Discuss how your comparison in Question 7 will be changed if both electrical melting and solar energy melt the ice. What other measures could be employed to reduce the solar/electrical costs? *(2)*

Data analysis 6 – Power storage systems (time allowed – 45 minutes)

This question considers the problem of storing energy as part of the national electricity system.

The demand for electrical energy in the UK varies throughout the day. Figure 9.5 shows a typical power against time graph for one 24-hour period.

➤

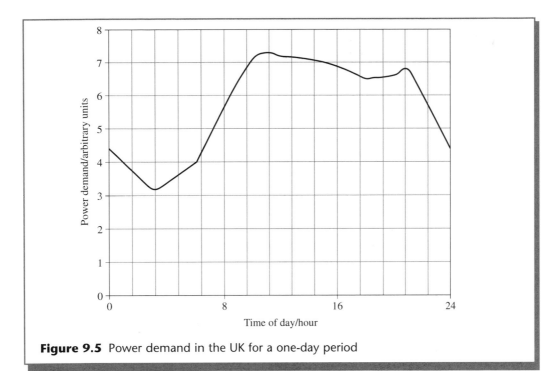

Figure 9.5 Power demand in the UK for a one-day period

1 One solution to the problem of meeting electrical demand is to build sufficient generating capacity to meet the peak demand. Estimate the number of hours per day for which at least one-quarter of the power generation system would not be used. *(2)*

2 An alternative solution is to provide two types of station. The first type supplies its power throughout the 24-hour period. The second type stores energy for part of the day and releases this energy at times of peak demand.

Use Figure 9.8 to comment *quantitatively* on the maximum power rating required for the storage systems compared with the first type of continuously operating system. *(3)*

3 A useful power for storage systems in the UK would be about 25 000 MW. Use Figure 9.8 to estimate the energy in kW-hour that the systems should be able to store. *(2)*

4 One possible design for a storage power station is to use off-peak energy to compress air into a sealed underground cave and then, during periods of high demand, to reverse the system so that the air can drive a generator.

One system of this type compresses air from atmospheric pressure (10^5 Pa) to 7 MPa in an underground cave of volume 3×10^5 m^3.

a) Calculate the volume of this air before compression.

b) Sketch a graph of pressure against volume for the air, assuming that the temperature is constant.

c) The area under the curve is given by the formula:
area = $P_1 V_1 \ln (P_2/P_1)$, where P_1, V_1 are the pressure and volume respectively before the compression and P_2 is the pressure after the compression.
Use the formula to estimate the stored energy in the air. *(7)*

5 Batteries of storage cells could also be used. Lead-acid batteries can store about 50 W-h kg^{-1} of their mass and generate a peak power of 70 W kg^{-1}.

a) Calculate the energy stored per kilogram mass of the battery.

b) Discuss with *quantitative estimates* the design of a battery system to meet the target in Question 3.

c) Figure 9.6 shows how cost per kW of the battery and air systems varies with daily discharge hours at their power rating. Explain why the cost versus hours relationships are different for the battery and the air systems. *(4)*

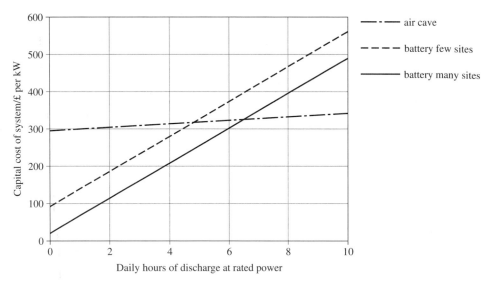

Figure 9.6

6 A third type of storage system pumps water at periods of low demand from an underground reservoir to a lake high in the mountains. At peak demand times the water is allowed to flow back down to the reservoir through a turbine and generating system. The stored water (density 1000 kg m^{-3}) drops through a vertical distance of 370 m as it does so.

a) Estimate the volume of water that is required to flow in order that this station will equal the energy storage capacity of the air system described in Question 4.

b) The pipe down which the water flows is 1.6 m in diameter. Estimate the speed of the water flow. *(4)*

ANSWERS

Chapter 3

Answer to 'Now try this' on page 33.

The obvious synoptic topic for capacitor discharge is the whole area of exponential decay, which is mentioned a number of times in this book. Perhaps the flow of water out of a burette could be an uncommon but straightforward example (the flow depends on the pressure, which in turn is related to the amount of water remaining in the burette). The cyclotron links up with deflection of charged particles in both magnetic and electric fields, and also with the possibility of one field cancelling out the other (as in the determination of e/m). The television tube is an excellent example here.

Wave theory can go in many directions, including radio, optical, sound and many other phenomena.

Chapter 5

Answer to 'Now try this' box on page 59.

1 If you assume a wall height of 3 m and a mass of 80 kg for the man with a time of 0.5 s to stop at the ground, then the rough speed at the ground is $(2gh)^{1/2}$, about 8 m s^{-1}.
$F = m(v-u)/t = 80 \times 8/0.5$, about 1300 N.

Answers to 'Now try this' box on page 59.

2 There are no formal answers to the estimates, methods you might try are:

1 Estimate the number of bricks (with a double wall thickness?) and then the weight of a brick.

2 Estimate the number of characters on one page and then the whole book, there were at least five drafts.

3 Estimate the area of the whole site and then decide on the fraction of site covered by school.

4 Have a look inside but remember that some notes have two or three strings.

Answer to 'Now try this' on page 63.

1 The resistance is $1.45/7.3 \times 10^{-3} = 198\ \Omega$, the fractional error in the resistance is $1/14 + 2/73 = 1/10.1$. So the final answer is (198 ± 20) ohm, or better, (200 ± 20) ohm.

2 $g = 4 \times \pi^2 \times 0.65/(1.616)^2 = 9.8263\ …$ Errors: 4 and π have no error, combining the others gives $1/650 + 1/404 = 1/294$, so error is $9.82/294$ and answer is 9.83 ± 0.04 m s^{-2}.

Answers to 'Now try this' on page 66.

2

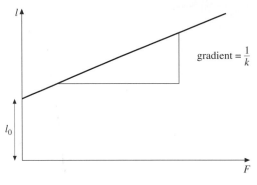

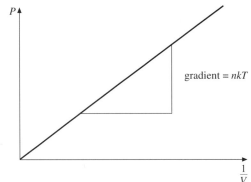

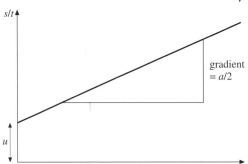

Figure 10.1

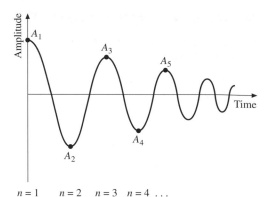

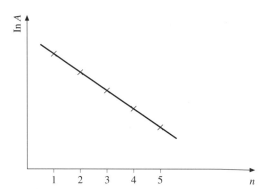

Figure 10.2

Answers to 'Now try this' on page 69.

(i) $y = 25x^{0.7}$

(ii) $y = 5.6x^2$

(iii) $y = 0.15x^{0.66}$.

Answer to 'Now try this' on page 68.

1 Plot Q against $\ln t$, the gradient will be $-1/RC$, the intercept on the Q axis will be Q_0.

Chapter 6

Answers to 'Now try this' on page 77.

1 $a = 1.76$ m s^{-2}; a against t is a straight line graph with constant y value from $t = 0$ s to $t = 15$ s; v against t, is a straight line from (0,0) to (15,21.6); s against t is a parabola from (0, 0) to (15, 198).

2 This graph should look like the $y = x^2$ graph with F on the y-axis.

Answer to 'Now try this' on page 77.

Check your sketches of the graphs against those in Chapter 6.

Answers to 'Now try this' on page 79.

$V = 5.2$ V; $I = 1.15$ A, so $R = 5.2/1.15 = 4.52$ Ω.

Answer to 'Now try this' on page 79.

$I = 0.05$ A; $V = 0.0406$ V; $I = 6$ A; $V = 49.5$ V. The smaller estimate is the most reliable.

Chapter 7

Answers to 'Now try this' on pages 99–103.

(d) (i) You will have to make a reasonable judgement of the absolute error. Perhaps ± 0.2 s might be reasonable for the time and ± 0.005 m for the depth of the water.

(ii) The fractional uncertainty is 1/70 in the 0.01 m depth result. Assume that the distance travelled is error free.

(e) a log–log graph is appropriate in this case. Your graph should lead to values of

$n = 0.5$; $k = 9.8$.

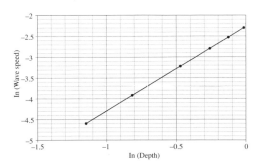

Figure 10.3

(f) The units of k are m s^{-2}.

2 **Emf induced in magnetic coils**

(e)

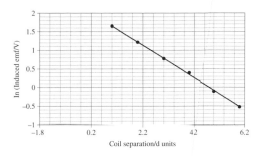

Figure 10.4

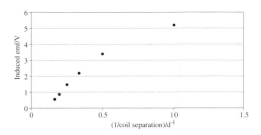

Figure 10.5

The second suggestion is correct.

(f) 7.8 V.

3 **Stressed wooden beam**

(d)

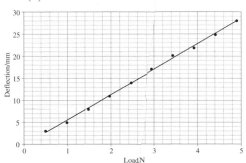

Figure 10.6

(e) The graph of deflection against load is a straight line. This is the behaviour that is expected by a material that obeys Hooke's law.

(f) $(2.3 \pm 0.4) \times 10^{10}$ Pa.

Chapter 9 –
Comprehension questions

Each answer from Chapter 9 is followed by ticks that show the number of marks assigned to it. Sometimes these ticks are within a solution to indicate where marks are gained for the method. You can use these ticks to mark your own work, but remember that examiners will give credit for working carried forward from earlier parts of the work. In this case you will need to check carefully to see that your *solution* is correct even though you may have begun with the incorrect *data*.

1 Orbiting generator

1 (a) (i) Gravitational pull of Earth. ✓

(ii) orbit radius = 6.6 Mm;
$v = (GM/r)^{1/2}$ ✓ = 7.8 km s^{-1} ✓

(b) Shuttle has orbit radius that is 20 km bigger than shuttle ✓, so its orbital time is slightly slower ✓.

(c) Either use calculus ($dv = \frac{-v}{2r}\,dr$) or a calculator to calculate speed to 6 significant figures at both radii ✓ ✓. Speed difference is 11.8 m s^{-1}. ✓

2 (a) Copper as a conductor ✓, Kevlar as strengthener ✓. **b)** Kevlar loses its strength and whole material relies on copper properties, so fails ✓.

3 (a) Electrons in wire constitute a current so magnetic force acts ✓ (similar to an electric motor). This force acts along the tether so electrons drift to one end ✓; there is an electron deficit at the other end ✓. Charge separation can do electrical work ✓ so there is an emf between ends of tether. ✓

(b) $\varepsilon = Bvd$ ✓;
$B = \dfrac{5000}{(8000 \times 20\,000)} = 31\ \mu T$ ✓
(c) Charge will build up until presence of charge is so large ✓ that further charge movement is prevented by repulsion ✓. Current flow in tether will stop.

(d) (i) $I = 1000/5000$ ✓ = 0.2 A ✓

(ii) 0.2 C s^{-1} ≡ 3.2 × 10^{20} electrons s^{-1} ✓

4 (a) Simple harmonic motion – the length and therefore the extension of 'spring' will vary sinusoidally ✓ so there will be sinusoidal variation of tension too ($F=kx$) ✓. You could sketch graph (Figure 10.5); it will be always +ve (above *x*-axis).

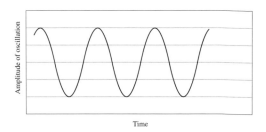

Figure 10.7

(b) Young modulus (YM) = stress/strain. Constant bobbing amplitude with shortening overall length means increased strain ✓. As YM is constant, this means increased stress ✓. Assuming cross-sectional area is constant, tension must increase ✓.

2 Weightless towers

1 (a) Situation in outer space ✓ well away from attraction of any objects with mass ✔.

(b) Simulated ✓; both space station and occupants fall towards Earth with same acceleration ✓.

2 (a) $s = 110$ m; $g = 9.8$ m s^{-2}; $u = 0$ m s^{-1}; $s = ut + \frac{1}{2}at^2$; $t = 4.7$ s ✓ ✓

(b) $v = u + at$; $v = 46$ m s^{-1}. ✓ ✓

(c) $v^2 = u^2 + 2as$; $a = 135$ m s^{-2} ✓ ✓

(d)

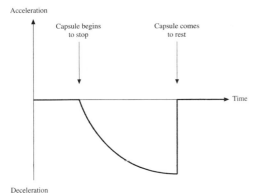

Figure 10.8

✓ ✓

3 (a) $D = 0.4 \times 46^2 \times 1.3 \times 10^{-5}$ ✓ $=$ 1.1×10^{-2} N ✓

(b) $g = F/m = \dfrac{1.1 \times 10^{-2}}{300}$ ✓ $= 3.7 \times$ 10^{-5} N kg^{-1} ✓

(c) Reaches terminal speed when $D = mg$ ✓; i.e. when $0.4\rho v^2 = mg$ ✓; $v = (300 \times 9.8/0.4 \times 1.3)^{\frac{1}{2}} = 75$ m s^{-1}, ✓ so: no ✓.

4 (a) Current in coil produces a *moving* magnetic field ✓, if this is moving relative to a stationary part of the tower (e.g. the walls), then an induced emf ✓ will be produced, leading to both a current and a magnetic field in the wall ✓. This will induce further currents in the capsule coil ✓and also decelerate the capsule ✓.

(b) $\varepsilon = Blv$ ✓ $= 70 \times 10^{-6} \times 0.4 \times 46$ $= 1.3$ mV ✓

3 Satellites for TV

1 (a) Orbit where satellite is stationary relative to Earth ✓; 24 hours ✓.

(b) (i) Equate the centripetal force to the gravitational force on the satellite and use the result $r^3 = GM_e/I\omega^2$ to calculate r. ✓ ✓
(ii) passage quotes distance from surface of Earth ✓.

(c) energy $= GM_e m/r = 1.7$ GJ ✓ ✓

(d) Geostationary satellites orbit over the Equator ✓. Gravitational force and required direction of centripetal force must be in same direction ✓. At any other latitude this is not the case. ✓

(e) Telstar in close Earth orbit ✓. Orbital time about 90 minutes, if in polar orbit access times even shorter. ✓

2 (a) $\lambda = c/f$ 25 mm ✓ ✓

(b) Microwave ✓

(c) $n\lambda = d \sin\theta$ ✓ yields $d = 1.75$ m ✓ with $n = 1$ ✓.

(d) 0.32 nW ✓ ✓

3 (a) 9.5 m^2 ✓ ✓

(b) Need to keep directed at Sun ✓, eclipsing by Earth ✓, stored energy is needed.

4 (a) 1.37×10^{-11} s^{-1} ✓ ✓

(b) Products are $_{-1}^{0}\beta$ ✓ $+ \, _{90}^{228}$Ac ✓ $+$ energy ✓.

(c) About 22 kg ✓ ✓ ✓ ✓

4 Inside a star

1 (a) Gravity ✓, (b) thermal pressure ✓.

2 Particles only interact when close together ✓. Most of the time (except during collisions) they do not affect each other. ✓

3 Electrons are stripped from atoms ✓. Forces between the electrons and ions have a much longer range ✓ than forces between neutral atoms in the other three states.

4 Effective distance over which force acts ✓. Sketch should show attractive $1/r^2$ dependence for both cases, ✓ gravitational case should be shown as very significantly smaller than electrostatic. ✓

5 Equate $p_{gas} = nkT$ with $p_{gas} = 1/3 \, \rho c^2$ ✓; C from M_{Sun}/V_{Sun} about 10^5 m s^{-1}. ✓

6 $p_{gas} = 5.5 \times 10^{13}$ Pa ✓; $p_{rad} = 4.0 \times 10^9$ Pa ✓. Confirms passage view that p_{rad} is of marginal importance in most stars. ✓

7 $p = nkT/\beta$; $p^3 = 3n^4k^4(1 - \beta)/\sigma\beta^4$ ✓ ✓

8 4.8×10^{26} W ✓ ✓

9 2.4×10^{-4} W kg^{-1} ✓ ✓

10 2.4×10^8 protons s^{-1} kg^{-1} ✓ ✓ ✓

11 0.75 kg of H nuclei ✓ $\equiv 4.4 \times 10^{26}$ nuclei ✓.

12 So 1.5×10^{26} ✓ fused at rate of 2.4×10^8 s^{-1}, this takes 6×10^{17} s ✓ ✓ $\equiv 20 \times 10^9$ years (an over estimate).

13 (a) $Q^2/4\pi\varepsilon_0 r$ ✓ $= 7.5 \times 10^{-14}$ J ✓. r (separation of proton centres) $= 3.0 \times 10^{-15}$ m.

(b) $kT = 7.5 \times 10^{-14}$ J ✓; thus $T = 5.4 \times 10^9$ K ✓

(c) Proton speeds are distributed from very high to low ✓. The very fast moving protons achieve fusion. ✓

14 It would be easier to obtain fusion from protons so many more fusions would occur at present temperature. ✓ So more power will be produced ✓ and temperature rises ✓. Radiated power and pressure will rise as T^4 ✓, whereas gas pressure rises as T. Radiated pressure will become dominant until new equilibrium arises ✓. Radiation from surface will increase, this will have major effect on life on Earth, both in terms of energy arriving ✓ and wavelengths of radiation. Sun's projected lifetime will decrease markedly. ✓ (5 ✓ max allowed)

5 Black holes

1 (a) Escape speed is speed such that kinetic energy at this speed ✓ is sufficient to overcome gravitational potential energy and allow the object just to reach an infinite distance with zero speed. ✓

(b) Equate $\frac{1}{2} mv^2$ to GMm/r. ✓ ✓

2 (a) 2.01×10^{30} kg ✓ ✓

(b) 6.2×10^5 m s^{-1} ✓ ✓

3 2.98 km ✓ ✓

4 $\frac{1}{2} mv_e^2 = 2.2 \times 10^{13}$ J ✓ ✓ ✓

5 2.6×10^{12} K ✓ ✓

6 (a) 1.43×10^{-14} m ✓ ✓

(b) Gamma ray ✓

7 Use $\Delta E = \Delta mc^2$; 0.512 MeV (for one of the two photons) ✓ ✓ ✓

6 Mobile telephone technology

1 No cables ✓; electromagnetic waves are invisible and inaudible ✓; no need to be able to see transmitter. ✓

2 Electromagnetic waves are diffracted ✓ and 'spread out and around' obstacles ✓.

3 $c = f\lambda$; 0.33 m ✓ ✓

4 1000 ✓ ✓

5 (a) Inverse-square law effects ✓

(b) $D = P/4\pi r^2$ ✓

(c) Absorption ✓, for example by nearby buildings or ground, or by water in the atmosphere. ✓

(d) 1.1×10^{-9} W ✓ ✓

6 (a) Minimum permissible distance between the outer edges of transmitting zones using the same frequency ✓.

(b) Maximum distance of mobile from required signal transmitter, strength of interfering signal that can be tolerated ✓.

(c)

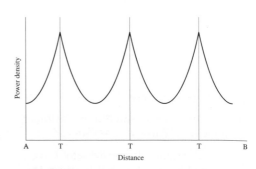

Figure 10.9

✓ ✓ ✓

(d) Worst case is 25 km from main cell, 75 km from interfering cell ✓, power ratio is $25^2/75^2 = 11\%$ ✓

Data analysis questions

1 Windmills in the sea

1 (a) Column of air length v cross-sectional area A hits turbine every second, density is ρ so mass is ρAv. ✓ ✓

(b) Energy arriving per second = kinetic energy per second = ½ (ρAv) v^2. ✓

2 For each bar on the graph determine the fractional contribution to ½ρAv^3 at the mean speed of the bar range (i.e there is a contribution of 0.33 × ½ρA × 0.3^3 from the first bar). ✓ ✓ ✓ ✓

3 Power = ½ρAv^3 = ½ $(\rho A)(CII^{0.15})^3$ ✓ ✓

(a) *Power* depends on *height*$^{0.45}$ so the higher the better. ✓

You could sketch Figure 10.10 to illustrate your answer.

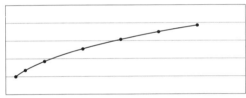

Height

Figure 10.10

(b) e.g. capital cost of structure. ✓

4 2.5×10^6 W = ½ × 1 × A × $(9 \times 1.5)^3$; $A = 2 \times 10^3$ m². ✓ ✓ ✓

5 This is a large area. It corresponds to a square 145 m on each side. ✓ The blade would be heavy ✓ and hard to fabricate. ✓

6 Air flow disturbances ✓, one turbine shielding another, ✓ etc.

7 $50 \times 600 \times 2500 \times 5/100 = £3.75$ million. ✓ ✓

8 (a) $R = \rho l/A$ ✓ = $1.7 \times 10^{-8} \times 60 \times 10^3/\pi(0.1)^2 = 0.032\ \Omega$ ✓.

(b) 125 MW at 132000 V ✓ so $I = 0.95$ kA. ✓

(c) $I^2R = 0.95^2 \times 10^6 \times 0.034$ ✓ = 31 kW ✓.

9 Weight = $9.8 \times 60\ 000 \times \pi \times (0.1)^2 \times 9000 = 1.7 \times 10^8$ N . ✓ ✓

10 (a) Al is less dense so lighter . ✓

(b) Al has small Young's modulus (YM), steel has much larger YM so maintains strength of cable. ✓

(c) Resistance of aluminium = 0.055 Ω ✓
Resistance of steel = 0.13 Ω ✓
Total resistance = 0.039 Ω . ✓

11 (a) $B = \mu_0 I/2\pi r = 4\pi \times 10^{-7} \times 950/2\pi \times 1.5 = 130$ μT ✓ ✓
$F = BIl = 130 \times 10^{-6} \times 950 \times 1 = 0.12$ N m^{-1}. ✓ ✓

(b) Currents always in opposite directions so repulsive. ✓

2 Power boat performance

1 (a) Power = force × speed ✓

Total mass of boat + cargo/kg	Power/W at a boat speed of		
	1m s^{-1}	2 ms^{-1}	3 m s^{-1}
500	200	580	1140
800	240	660	1380
1100	290	800	1650
1400	350	960	1980
1700	430	1200	2490

Table 10.1

✓ ✓

(b)

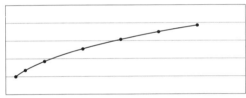

Figure 10.11 Power plotted against mass at three speeds

Synoptic Skills in Advanced Physics

2 (a) Total mass = 1000 kg; read-off gives 750 W. ✓

(b) 2.2 kW at 3 m s^{-1}; so mass = 1550 kg, i.e. 1050 kg of cargo. ✓

(c) 1.5 kW, total mass 1500 kg; interpolation gives 2.5 m s^{-1} ✓

3

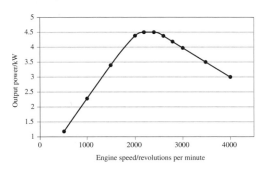

Figure 10.12

✓ ✓ ✓

4 (a) System A needs 660 W from 4.4 kW so ratio = 6.7:1. ✓

System B needs 660 kW from 4.0 kW so ratio = 6.1:1. ✓

(b) 1 litre of fuel gives 40 MJ which returns 8 MJ of energy after 20% conversion. ✓

System A converts this to 8 MJ/6.7 = 1.19 MJ hence 1.19/40 × 100 = 3.0% ✓

System B converts this to 8MJ/6.1 = 1.31 MJ hence 1.31 /40 × 100 = 3.3% ✓

(c) System A; 1 litre yields 1.19 MJ; work done = force × distance; so 3.6 km; 10 litres gives 36 km. ✓

System B; 4.0 km; 10 litre gives 40 km . ✓

(d) System A; 80 p/litre so 1 km costs 22 p. ✓

System B; 1 km costs 20 p. ✓

3 Ozone in the atmosphere

1 (a) A suitable graph would be ln(P) against h; a straight line, ✓ with negative gradient, ✓ .

(b) The constant-ratio test is best: a *change* in height of 20 km always gives a ratio of pressures of $P1/P2 = 10$, ✓ carry out test twice. ✓

(c) Equation leads to $\ln(P) = \ln(P_0) - ah$, so $\ln(1) = \ln(100) - a \times 4 \times 10^4$ ✓; so $a = 1.15 \times 10^{-4}$ m^{-1}. ✓

(d) $P = 100e^{-0.000115 \times 9000}$; $P = 0.355$ ✓; 64.5% of atmosphere is *below* the climber. ✓

2 (a) Use $pV = NRT = \frac{1}{3}\rho$ (*mean square speed*) ✓; root mean square speed = 1670 m s^{-1}. ✓

(b) Escape speed = $(2GM/r)^{\frac{1}{2}}$ ✓; substitute to get 11.2 km s^{-1} . ✓

(c) Small but significant number of atoms have very high speeds ✓ well above root mean square value; if the speed exceeds escape speed, then they escape. ✓

3 (a) (i) Either plot two graphs on one set of axes or plot the *difference* between values ✓ ✓ ✓

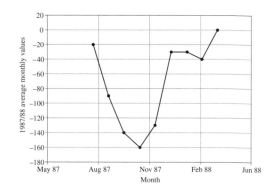

Figure 10.13

(ii) Dramatic falls in ozone levels ✓, by about one-third in value, so less protection from *uv* rays. ✓

(b) 1 DU is amount of ozone in layer of ozone 10^{-5} m thick and 1 m^2 in area.

So: $1 \times 10^{-5} \times$ (density of ozone = 2.1)=21 mg. ✓

(c) Maximum change in DU totals is between Sept and Dec = about 240 DU (Aug/Dec) to 140 DU in Oct, so 100 DU loss; Antarctic area is 10^{13} m^2.

Therefore, mass = $100 \times 10^{13} \times 20 \times 10^{-6}$ = about 2×10^{10} kg. ✓ Ozone concentrations are constant over the whole Antartic area. ✓

(d) This mass would require over 2 million flights per year, i.e. 6000 per day ✓. Need to manufacture such a large mass, fuel/transport costs very high. ✓

4 Froude numbers

1 (a) $(60 \times 9.8)/2 \times 5 \times 10^{-4} = 5.9 \times 10^5$ Pa. ✓ ✓

(b) (i) Area goes up $\times 4$ (2^2) ✓.

(ii) Volume goes up $\times 8$ (2^3) ✓.

(iii) Density is constant so weight goes up $\times 8$ too ✓.

(c) Force (weight) goes up $\times 8$; area goes up $\times 4$ so stress goes up $\times 2$. ✓

(d) (i) Acceleration upwards requires force so an additional force acts on bone; total force increases. ✓

(ii) Peak force = $3.5 \times 60 \times 9.8$; stress = $3.5 \times 60 \times 9.8/5 \times 10^{-4}$ ✓ = 4.1×10^6 Pa. ✓
(iii) Breaking stress (15×10^6 Pa) is $3.6 \times$ peak force, ✓ so greatest possible height is $1.7 \times 3.6 = 6.1$ m. ✓

2 (a) (i) $F = v^2/gl = 1.5^2/(9.8 \times 0.8) = 0.29$. ✓

(ii) $F = 9^2/(9.8 \times 0.8) = 10.3$. ✓

(b) (i) $0.8 \times 3.6 = 2.9$ m ✓.

(ii) Froude number is constant at 10, ✓ so $v^2 = 10 \times 9.8 \times 2.9$; $v = 16.9$ m s^{-1}. ✓

(c) (i) v^2/gl increases as v increases; s/l increases as v^2/gl increases; l constant so s increases. ✓

(ii) $F_{walking}$ for present human = 2.9, so $s/l = 1.6$; $s = 1.6 \times 0.8 = 1.28$ m. ✓ ✓

$F_{sprinting}$ for present human = 10, so $s/l = 5.2$; $s = 4.2$ m. Ratio is 3.3 : 1. ✓ ✓

(d) Enlarged human leg length = 2.9 m, so $s/l = 5.2$; $s = 14.6$ m. ✓ ✓

3 (a) (i) $v = 9$ m s^{-1}, $m = 60$ kg, ke = $\frac{1}{2} \times 60 \times 9^2 = 2.4$ kJ. ✓ ✓

(ii) $v = 17$ m s^{-1}, $m = 60 \times 8$, ke = $\frac{1}{2} \times 480 \times 17^2 = 69$ kJ. ✓

(b) Providing a sufficiently large rate of energy conversion might be a problem. ✓

5 Towing icebergs

1 About 7×10^8 m^3 every year. ✓ ✓

2 7 or 8 small ones; 1 large one ✓ assuming no losses on route. ✓

3 Large one needs one tow per year and is overall cheaper by a factor of 4 ✓; smaller ones can probably go faster (losing less overall from melting) and possibly more reliable. ✓

4 250 m thick, so area is 4×10^5 m^2, which accumulates $600 \times 4 \times 10^5 = 2.4 \times 10^8$ J every second. ✓ Journey takes approximately $10^7/0.8$ s = 1.25×10^7 s, ✓ so 3×10^{15} J available for melting, this melts $\dfrac{3 \times 10^{15}}{3.4 \times 10^5}$, about 10^{10} kg. ✓ Mass of entire iceberg is $10^8 \times$ density (1000 kg m^{-3}) = 10^{11} kg, ✓so about 90% survives the journey. ✓ (assumes no reflection.)

5 Energy required to melt iceberg = $10^{11} \times 3.4 \times 10^5$ J = 3×10^{16} J. ✓ Energy to tow = force $\times$ distance = $5 \times 10^6 \times 10^7 = 5 \times 10^{13}$ J. ✓ So towing energy is about 600 times less. ✓

6 With $4 \times 10^5 \times 600$ J s^{-1} arriving, it will require $\dfrac{3 \times 10^{16}}{2.4 \times 10^8}$ ✓, about 1.2×10^8 s ✓ to melt the ice in sunlight only, at least 3.8 ✓ years.

The 250 MW generating station will deliver energy at the same rate ✓. So the time will be constant. The sunlight is free, but the energy supply will decrease as the area shrinks. ✓ Also, the iceberg (which is 0.7 km on each side) is likely to create a microclimate with its own fog shield!

7 3.6×10^6 J (1 kW-hour) costs £0.08. One day of electrically produced water costs $2 \times 10^9 \times 3.4 \times 10^5 \times 0.08/3.6 \times 10^6 = £16$ million. ✓ One day of desalinated water costs $2 \times 10^6 \times 1.5 = £3$ million. ✓

8 With the assumption that solar and electrical energy are matched joule for joule, then this will reduce the daily cost to £8 million ✓. Additional features could be used to increase the contribution from solar energy, e.g. focussing mirrors, etc. This would reduce the cost even further towards a break-even point with the desalination plant. ✓

6 Power storage systems

1 Peak demand requires 7.3 power units, minimum is 3.3 power units, 'top' 25 % is 6.3–7.3. Demand is less than 6.3 for about 11 hours in 24 hours. ✓ ✓

2 Area under graph is energy consumption. ✓ Count squares between max and min values, about 14 squares. So half of this (7) is the energy that the continuous stations must make for storage (assuming 100% efficiency). ✓ Leads to value of about 5.3 power units for continuous output. ✓

3 Storage would be generating for 15 hours, storing for 9 hours from graph. So, $25 \times 10^6/9$ kW-hour $= 2.8 \times 10^6$ kW-hour. ✓ ✓

4 (a) pV = constant; $V = 2.1 \times 10^7$ m^3. ✓ ✓

(b)

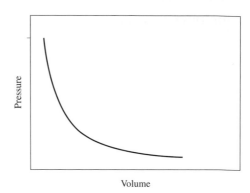

Figure 10.14

(c) 8.9×10^{12} J. ✓ ✓ ✓

5 (a) $50 \times 3600 = 1.8 \times 10^5$ J. ✓ ✓

(b) 49 million kg = 49 000 tonnes of cells are required. ✓

(c) Air system cost is likely to be dominated by cave creation. ✓

6 (a) Use $mgh = gpe$; $8.9 \times 10^{12} = 1000 \times V \times 9.8 \times 370$; $V = 2.5 \times 10^6$ m^3. ✓ ✓

(b) $4 \times 10^8 = 1000 \times V \times \pi \times (0.8)^2$. $V = 55$ m s^{-1}. ✓ ✓